S0-CAH-298

PROGRESS IN BIOORGANIC CHEMISTRY

PROGRESS IN
BIOORGANIC CHEMISTRY Series

VOLUME THREE

Edited by

E. T. KAISER

Departments of Chemistry and Biochemistry
University of Chicago

F. J. KÉZDY

Department of Biochemistry
University of Chicago

A WILEY–INTERSCIENCE PUBLICATION

JOHN WILEY & SONS

NEW YORK · LONDON · SYDNEY · TORONTO

SEP

clem.

Copyright © 1974, by John Wiley & Sons, Inc.

All rights reserved. Published simultaneously in Canada.

No part of this book may be reproduced by any means, nor transmitted, nor translated into a machine language without the written permission of the publisher.

Library of Congress Cataloging in Publication Data:

Main entry under title:

Progress in bioorganic chemistry.

(Progress in biorganic chemistry series)
Includes bibliographies.
1. Biological chemistry—Collected works. 2. Chemistry, Physical organic—Collected works. I. Kaiser, Emil Thomas, 1938–   ed. II. Kézdy, F. J., 1929–   ed.

QD415.A1P76   547'.1'3   75-142715
ISBN 0-471-45487-7 (V. 3)

Printed in the United States of America

10 9 8 7 6 5 4 3 2 1

P: SD· 12/17/24

QD 415
A1P76
v. 3
CHEMISTRY
LIBRARY

CONTRIBUTORS

LAWRENCE J. BERLINER, *Department of Chemistry, The Ohio State University, Columbus, Ohio*

ROBERT L. HEINRIKSON, *Department of Biochemistry, University of Chicago, Chicago, Illinois*

KARL J. KRAMER, *Department of Biochemistry, University of Chicago, Chicago, Illinois*

M. LAZDUNSKI, *Centre de Biochimie, Université de Nice, Nice, France*

01908

FROM THE
PREFACE TO THE SERIES

Bioorganic chemistry is a new discipline emerging from the interaction of biochemistry and physical organic chemistry. Its origins can be traced to the enzymologists whose curiosity was not satisfied with the purification and the superficial characterization of an enzyme, to the physical organic chemists who had the conviction that the elementary steps of biological reactions are identical to those observed in organic chemistry, and to the physical and organic chemists who wished to understand and to imitate *in vitro* the unequaled catalytic power and specificity exhibited by living organisms. As with all interdisciplinary sciences, bioorganic chemistry uses many of the methods and techniques of the disciplines from which it is derived; many of its protagonists qualify themselves as physical organic chemists, enzymologists, biochemists, or kineticists. It is, however, a new science of its own by the criterion of having developed its own goals, concepts, and methods.

The principal goal of bioorganic chemistry can be defined as the understanding of biological reactions at the level of organic reaction mechanisms, that is, the identification of the basic parameters which govern these reactions, the formulation of quantitative theories describing them, and the elucidation of the relationships between the reactivity and the structures of the molecules participating in the process. This definition is narrower than one which some scientists would give. They might prefer to include areas such as medicinal chemistry, for example, as part of the field of bioorganic chemistry. Accordingly, the goals which they would cite would differ from those which we have considered. We do not seek here to argue or to defend our concept of what constitutes bioorganic chemistry, but within the framework of our definition we believe that there is a real distinction between much work in present day medicinal chemistry and that in the bioorganic field. In our conception of bioorganic chemistry the emphasis is on mechanism.

The theoretical formulation of the understanding of biochemical and, therefore, enzyme-catalyzed reactions has required the elaboration of new concepts, such as multifunctional catalysis, stereospecificity by three-point attachment, and control of reactivity by conformational changes. Many of the new concepts will not survive; they will be redefined, discarded, or reevaluated as fortunately always happens in science. But the trend is clearly apparent—these new concepts are providing us with efficient tools of great power which can be used to describe and discuss enzymatic reactions.

As to the methods of bioorganic chemistry, they are conceptually the same as those for the study of any chemical reaction; they include analytical and physical techniques. However, the complexity of the reacting molecules has resulted in methods which are new and unique in their ability to probe the chemistry of a functional group surrounded by a multitude of very similar groups or a chemical event accompanied by a host of satellite reactions. The discovery of numerous methods involving active-site directed reagents, "reporter molecules," and chromophoric substrates illustrates the usefulness and the elegance of the new science.

The future of bioorganic chemistry appears very promising, and the fields to cover in the future are immense and unexplored. The earliest work has concentrated on the understanding of general acid—general base catalyzed reactions, hydrolytic reactions, and the role of proteins in enzymatic catalysis. The mechanism of enzymatic catalysis by most coenzymes is very far from being well described, and the very prominent role of metal ions in catalysis is only beginning to emerge. Other important problems, such as surface catalysis at biological membranes, transport mechanisms, the process by which ribosome-catalyzed reactions occur, and the reactivity of

RNA and DNA molecules, are at an early stage of development or are completely unexplored.

As a result, a rapid growth of bioorganic chemistry is desirable and is currently underway, as evidenced by the large number of papers published on the subject. An unfortunate result of this rapid expansion is the scarcity of comprehensive treatments of bioorganic chemistry. The rapid progress in this field makes it likely that large portions of any comprehensive textbook will become obsolete soon, although the student of bioorganic chemistry may still learn some of the basic concepts of the subject from them. Because of many factors, it would be possible to revise textbooks only at infrequent intervals. For this reason the format of presenting comprehensive treatments of limited subjects seemed more appropriate to us. It would provide the investigators, interested readers, and students with a thorough and critical evaluation of those aspects of bioorganic chemistry where definite and substantial progress has been achieved.

It is the hope of the Editors of this series to be able to respond to the need for up-to-date comprehensive treatments of important topics in bioorganic chemistry. In attempting to do so we would like to provide treatments of bioorganic subjects which will be general enough to retain the attention of most workers in the field and which will be at a level beyond that of a usual review article or literature survey. Since many aspects of bioorganic chemistry are still in the process of evolution, we also would like to provide a forum where the authors can express challenging new ideas and present stimulating and, frequently, controversial discussions. For this reason we hope to give the authors somewhat more latitude than is customary in this kind of publication, while still retaining the requirement of scientific sobriety.

E. T. KAISER
F. J. KÉZDY

Chicago, Illinois
January 1971

PREFACE TO VOLUME THREE

The present volume is the third in the series dedicated to summarizing progress in bioorganic chemistry. The reader will note that, as in the preceding two volumes, the emphasis is on physical, chemical, and kinetic methods used in studying enzymes and proteins. We feel strongly that the series should not be exclusively devoted to this aspect of bioorganic chemistry. However, this choice of topics in the early volumes is intended as a testimonial to the widespread use of these powerful analytical methods. The study of the structure-function relationship in enzymatic reactions is undeniably one of the focuses of interest of present day bioorganic chemistry.

The chapters in this volume should convey some sense of achievement with regard to the accelerating rate at which enzymes are yielding their secrets as new methods are elaborated and new approaches are explored. If the reader finds here some of the enthusiasm and the excitement that always accompanies new discoveries, the labor of the authors is well rewarded.

Chicago, Illinois   E. T. KAISER
January 28, 1974   F. J. KÉZDY

CONTENTS

APPLICATIONS OF SPIN LABELING TO STRUCTURE-CONFORMATION STUDIES OF ENZYMES

LAWRENCE J. BERLINER

Department of Chemistry
The Ohio State University
Columbus, Ohio

1

1 INTRODUCTION–HISTORY, PHILOSOPHY, AND PSYCHOLOGY OF SPIN LABELS

Several years ago, Koshland proposed a general method for probing biological structures for which he coined the term "reporter group" [1]. An illustration of the reporter group approach is depicted in Figure 1. The basic scheme involves the introduction of an environmentally sensitive chemical group into specific regions or sites of, for example, a protein molecule. This environmentally sensitive reporter group must possess some physical property that enables it to (1) "report" changes in its environment to an appropriate detection device, and (2) be distinct from the rest of the protein and thus suffer from little or no "background" signal from the protein itself. This is accomplished by choosing a reporter group with a physical property sufficiently different from that of the protein. Last, the probe must not add any significant perturbations to the biomolecular conformation or to other (catalytic) properties. Some physical techniques and properties that might be appropriate to a reporter group study are listed in Table 1. Thus a reporter group molecule strategically placed in a specific region of a biomolecule may provide a wealth of physical information about its environment. One of Koshland's principal points about the need for such methodology

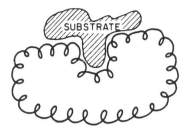
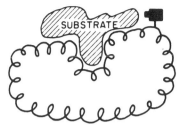

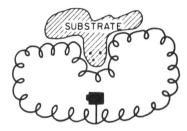

Figure 1 Schematic representation of enzyme-substrate complex in native protein, protein containing reporter group (solid area) adjacent to substrate binding area and reporter group at a distance from substrate binding area. From reference 1 with permission.

was with respect to the growing awareness that conformational or structural changes in proteins may be coupled to enzyme action. He had recently proposed that the traditional view of an enzyme-substrate reaction as a "lock-and-key" model was inconsistent with the high catalytic efficiency of several enzymes toward a variety of structurally different substrates [3]. His theory proposed that an enzyme must be flexible enough to accommodate each different substrate molecule in a highly productive complex for efficient catalysis. That is, an enzyme-substrate complex should result from an "induced fit" of substrate to enzyme [3]. Consequently, an ideal reporter group would be one that could report not only the static physical properties of an enzyme structure, but conceivably its dynamic properties as well.

The spin label method offered several advantages [4]. Since it was shown that the electron spin resonance (esr) of nitroxide spin labels was sensitive to molecular motion, orientation, and electric and magnetic environment, it potentially offered the ability to probe all these features in a biomolecular environment. In practice, as will be seen, the method has been found to be most sensitive to the first two

TABLE 1 A SELECTION OF REPORTER GROUP TECHNIQUES AND THEIR SENSITIVITY TO THE PHYSICAL ENVIRONMENT

Technique	Physical Method	Principal Sensitivity to Environment
	Electronic	
UV-visible	Optical absorption spectrophotometry	Polarity
CD	Polarized absorption spectrophotometry	Chemical, electronic environment
Fluorescence	Emission spectrophotometry	Polarity; some specific aspects of the chemical environment (proximity to other fluorescence acceptors or donors)
Fluorescence depolarization	Polarized emission spectrophotometry	Motion; polarity, chemical environment (as above)
	Magnetic	
Nmr	Nmr spectroscopy (absorption or pulsed)	Magnetic environment, motion, polarity
Esr (spin labeling)	Esr spectroscopy	Motion, orientation, polarity, magnetic environment
	Other	
γ-Ray Perturbed-angle correlation spectroscopy [2]	Radioactive coincidence spectrometry	Motion

properties: motion and orientation. That is, a small paramagnetic reporter group (spin label) incorporated at a specific site in a protein can relay information about the structural nature and (conformational) changes in its physical environment from an analysis of its motional freedom.

Historically, the evolution of the method required the discovery of a general class of stable versatile *paramagnetic* compounds. Unfortunately, most organic free radicals are unstable in biological media (water), and, on the other side, inorganic transition metal ions are limited in versatility. Although the very first spin label experiment of Ohnishi and McConnell [5] did not utilize the common spin label group, it was ultimately the successful synthesis of a nitroxide by Hoffman and Henderson [6], and not without the excellent development of the chemistry in this area by the very prolific groups of Rozantsev [7, 8] and Rassat [9], that signaled the birth of applications of this technique by McConnell [4, 10, 11].

2 PHYSICAL BACKGROUND—WHAT CONTRIBUTES TO THE ESR SPECTRUM?

2.1 Theory—Motional Effects

The nitroxide moiety is depicted in Figure 2. It contains an unpaired electron which is localized on the nitrogen and oxygen atoms. The flanking methyl groups are absolutely necessary to impart stability to the free radical, unless the nitroxide is attached to bridgehead carbons [48] (see Section 3.1).

The spin Hamiltonian $\hat{\mathcal{H}}$ describing a paramagnetic molecule such as a nitroxide is given by

$$\hat{\mathcal{H}} = \beta \, \mathbf{H} \cdot \mathbf{g} \cdot \hat{\mathbf{S}} + \hat{\mathbf{S}} \cdot \mathbf{A} \cdot \hat{\mathbf{I}} - g_n \beta_n \hat{\mathbf{I}} \cdot \mathbf{H} + \text{electron-electron dipole terms}$$
$$+ \text{electron-electron exchange terms} \quad (1)$$

where β = electron Bohr magneton, $\mathbf{H}$ = applied magnetic field, $\mathbf{g}$ = g-value tensor for the electron, $\hat{\mathbf{S}}$ = spin operator for the electron, $\mathbf{A}$ = electron nuclear hyperfine tensor (also designated T), $\hat{\mathbf{I}}$ = spin operator for a coupled nuclear spin (^{14}N), g_n = g factor for a nuclear spin (^{14}N), β_n = nuclear Bohr magneton.

In most applications with nitroxides, only the first two terms are significant. The third term is the spin Hamiltonian for the nucleus, which is included for generality but is usually unimportant. The last

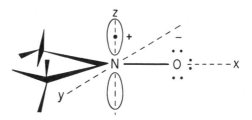

Figure 2   Structure of the nitroxide moiety containing the free electron in the $2p_z$ orbital of the nitrogen atom. The principal axis system x, y, z of the spin Hamiltonian for a nitroxide radical is also designated. From reference 54 with permission.

two terms are significant only in very special cases (discussed later). The energy level diagram and resultant esr spectrum described by this spin Hamiltonian are shown in Figure 3. Thus the spectrum of a free electron (spin $S = \frac{1}{2}$) coupled to a ^{14}N nucleus (nuclear spin $I = 1$) consists of three lines $(2I + 1)$, equally spaced by a hyperfine coupling constant A, somewhat analogous to the nuclear magnetic resonance (nmr) case of a proton split by two neighboring equivalent protons. The spectrum in Figure 3 also represents that of a small nitroxide molecule tumbling rapidly in solution. In this special case

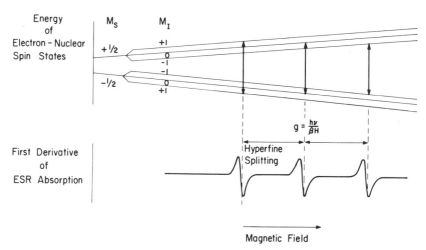

Figure 3   The esr experiment for a species containing an unpaired electron of spin $\frac{1}{2}$ and a nucleus (^{14}N) of spin 1. The separations between the resonance lines yield the hyperfine splitting constant. From reference 148 with permission.

essentially all the nitroxide molecules in the macroscopic sample have the same spin Hamiltonian,

$$\hat{\mathcal{R}} = g_0 |\beta| H \hat{S}_z + A_0 \hat{S} \cdot \hat{I} \tag{2}$$

where g_0 and A_0 are the isotropic components of g and A. The first term may be evaluated from the relationship $h\nu = g_0 \beta H$, where h is Planck's constant and ν is the esr spectrometer frequency. For nitroxides g is slightly greater than 2.00, corresponding to, in the most common (X-band) spectrometers, $\nu = 9500$ MHz (9.5 GHz) and H = 3400 G.

The spin Hamiltonian, however, tells us that in general both the g value (first term of eq. 1) and hyperfine coupling or splitting (second term of eq. 1) are dependent upon the *orientation* the N−O groups take with respect to the applied magnetic field direction. This *anisotropy* manifests itself in the following manner: A rigidly oriented nitroxide group (for instance, when substituted in a host crystal) exhibits a three-line esr spectrum with hyperfine coupling constants of from 5–8 G to 31–34 G. The g value, likewise, may vary from 2.0022 to 2.0088. For example, for di-t-butyl nitroxide in tetramethyl-1,3-cyclobutanedione host crystals, the ^{14}N hyperfine and electron g-tensor elements found were $A_{xx} = 7.6$ G, $A_{yy} = 6.0$ G, $A_{zz} = 31.8$ G; $g_{xx} = 2.0088$, $g_{yy} = 2.0062$, and $g_{zz} = 2.0027$ [12]. Figure 2 shows the principal molecular axes x, y, z of the respective *principal* values of the hyperfine and g tensor for the nitroxide group.

As discussed above, spectral *anisotropy* is observed for an oriented crystalline nitroxide, and spectral *isotropy* is observed for a free, rapidly tumbling nitroxide in solution. But what happens in between these two extremes of motion and order? This is exemplified by the esr spectra shown in Figure 4 for four different nitroxide radicals. They all have quite similar A_0 and g_0 values given by

$$A_0 = \frac{1}{3}(A_{xx} + A_{yy} + A_{zz}) \quad \text{and} \quad g_0 = \frac{1}{3}(g_{xx} + g_{yy} + g_{zz}) \tag{3}$$

where the terms on the right-hand side of each equation are the *principal* hyperfine and g-tensor values found from studies of these nitroxides in host crystals. As the viscosity of the water solution is increased by adding glycerol and lowering the temperature, the individual lines apparently broaden independently. At the very slow motional extreme of a frozen rigid glass, a very broad spectrum is observed which is 62–66 G in entire width (bottom spectra, Figure 4).

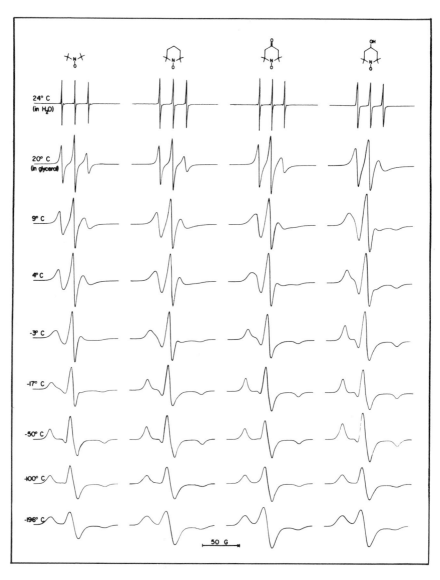

Figure 4 Effects of viscosity on nitroxide esr spectra. The top row of spectra represent a 10^{-4} M aqueous solution at $24°C$ of the spin label shown above them. All other spectra were recorded at spin label concentrations of 5×10^{-4} M in reagent grade glycerol (Mallinckrodt, $> 95\%$ by volume). Each column then represents the spectrum of the spin label above at the temperatures designated in the left-hand column. All spectra were taken at X band (9500 GHz). From reference 51 with permission.

This rigid glass or "powder" spectrum is observed when the molecular tumbling rate of the spin label is slower than the "limits of detection" of the esr experiment, which at X band (9.5 GHz) is about 10^8 sec^{-1}. In comparing macromolecular tumbling times, we find that most globular proteins are tumbling at rates below or near this limit. Thus a spin-labeled protein displaying a rigid glass or powder spectrum is strongly indicative of a nitroxide bound rigidly to a protein molecule that is tumbling slower than the lower limit discussed above. For lack of better terminology, spin label spectra in the fast-tumbling range (top third of Figure 4) have been designated weakly immobilized, spectra in the slow-tumbling range (lower third of Figure 4) have been designated strongly immobilized, and spectra in the intermediate range, moderately immobilized or of intermediate immobilization [4, 10].

A further aspect that should be at least mentioned here is whether the spin label is undergoing isotropic or anisotropic motion while on the macromolecule. Anisotropic motion (preferred rotation about one of the principal axes in Figure 2) can lead to complex line shapes which, in general, differ significantly from those isotropic spectra in Figure 4. This problem has been treated extensively for spin label probes in membranes or liquid crystals, in both of which anisotropic motion is significant [11, 13, 14]. However, in the case of proteins, this problem has not been examined very carefully, since (1) most of the spectra obtained have line shapes closely approximating those for isotropic motion, and (2) in the very slow tumbling rate region it is usually assumed that the isotropically (?) tumbling protein molecule dominates the motional characteristics of the bound spin label. It should also be pointed out that historically the anisotropic tumbling problem has only recently become of interest to theoretical analysis, and specifically with respect to membrane studies.

For the motional tumbling range represented by the spectra in the top half of Figure 4, a theoretical relationship may be derived relating the rotational correlation time τ_c to the line shape [4]:

$$\frac{T_2(O)}{T_2(M)} = 1 - 4 \frac{\tau_c}{15} b \Delta\gamma H_0 T_2(O)M + \frac{\tau_c}{8} b^2 T_2(O)M^2 \qquad (4)$$

where $T_2(O)/T_2(M)$ = ratio of the line width of the center line to that of the low field ($M = +1$) or high field ($M = -1$) line, τ_c = rotational correlation time for isotropic molecular tumbling, M = nuclear spin (^{14}N) quantum number, H_0 = applied field, b, $\Delta\gamma$ =

constants related to the hyperfine and g-tensor anisotropy of the particular nitroxide under study.

The line width ratio $T_2(O)/T_2(M)$ may be approximated by the peak-to-peak height ratio $\sqrt{h(M)/h(O)}$ if one assumes that the area under an esr line (derivative curve) is proportional to its height times width squared. It is important to emphasize that this calculation is most accurate for calculating *relative* rather than absolute rotational correlation times.

Attempts have been made to obtain τ_c values from a simple Stokes' law calculation for a spin label tumbling in a viscous medium [15, 16]

$$\tau_c = \frac{4\pi\eta r^3}{3kT} \tag{5}$$

where η = solvent viscosity, r = molecular radius, k = Boltzmann's constant, T = absolute temperature.

The differences between some of the spectra of Figure 4 under identical conditions illustrate how spin labels of comparable molecular size tumble at *different* rates at the same solvent viscosity. Nevertheless, comparison of these empirically derived spectra (Figure 4) with those for spin-labeled macromolecules still offers a valuable qualitative measure of spin label motion.

For the slow motional tumbling range (spectra in the bottom half of Figure 4), both theoretical and semiempirical treatments have recently become available [17, 18, 19]. Although a discussion of these techniques is beyond the intended scope of this section, a rotational correlation time estimate for an experimental spectrum involves comparison with computer-simulated spectra.

What about the case mentioned earlier in which a protein molecule is tumbling *near* but not significantly below the lower tumbling time limit? In fact, the observed tumbling rate of the spin label would be some complex function of the nitroxide-protein and whole-protein rotational diffusion tumbling rates. It is very important, especially in comparative spin label structural studies of homologous proteins, to normalize results by eliminating this latter tumbling contribution. However, macromolecular rotational correlation times are difficult parameters to measure accurately. Therefore it would also be valuable if one could measure this latter parameter as well. Shimshick, McCalley, and McConnell recently reported a relatively simple technique for obtaining protein rotational diffusion rates [20, 21]. A strongly immobilized spin-labeled protein was studied in increasing viscosities of sucrose solution, and the hyperfine extrema of the paramagnetic resonance spectrum were measured and extrapolated to infinite viscosity. Theoretically, it was noted that the quantity

$$\Delta H(\tau_2) = H_\infty - H_\eta \qquad (6)$$

is proportional to $\tau_2^{-2/3}$ for a wide range of correlation times, where H_η = position of the high-field component of the spectrum (see Fig. 4, spectra below -17°C) at viscosity η; H_∞ = position of the high-field component for a solution at infinite viscosity; and τ_2 = rotational diffusion correlation time of the protein. In practice then, one plots the shift in this hyperfine extremum versus $(T/\eta)^{2/3}$, extrapolates to infinite viscosity, and fits $\Delta H(\tau_2)$ to a theoretically calculated plot to find τ_2 for the spin-labeled protein in a solution of viscosity η. An experiment with α-chymotrypsin yielded a rotational correlation time of 12 ± 2 nsec in aqueous solution at 20°C, compared with 16 nsec for a similar measurement by fluorescence depolarization techniques [22]. There is also no doubt that *most* spin labels undergo anisotropic motion when incorporated in biological systems. Certainly, with the sophistication developed in simulating spectra of anisotropic tumbling motion in membrane systems, it should be possible in the future to detect, and perhaps quantitate, the extent of anisotropic tumbling motion of spin labels in proteins [23, 24, 97]. Such information should allow us to draw more detailed conclusions about the geometry of the structural environment of the spin label.

2.2 Other Effects

Both *g*-factor and hyperfine coupling are dependent on solvent polarity [25, 9]. For instance, the isotropic hyperfine coupling constant (line separation) for the nitroxide in the far upper-left spectrum of Figure 4 is 17.16 G in water and 15.10 G in hexane [23]. The isotropic *g*-factor changes from 2.0056 to 2.0061, respectively [9]. However, both of these effects are usually less important (and difficult to discern) in the broader line spectra corresponding to decreased tumbling rate (bottom half of Figure 4), as frequently found in protein studies.

When studying spin labels at concentrations exceeding 10^{-3} M or in cases in which two N–O groups are in close proximity (~ 10 Å) [such as in molecules LVI (Fig. 5) or those constraints dictated by a special protein environment], the two additional *electron-electron* terms in the spin Hamiltonian (eq. 1) become important. In the simplest case, high concentrations ($>10^{-3}$ M) of rapidly tumbling nitroxide radicals are distinguished by a symmetrical broadening of the typical three-line spectrum. As two (or more) nitroxide groups come in close proximity, the line shape becomes quite complex, depending upon the "internitroxide" distance and the relative tumbling rate of each radical. These phenomena have been important in spin-labeled proteins in only a few special cases, some of which are covered later in this chapter. The reader is referred to several texts on esr for a more detailed theoretical background [16, 26-29, 97].

3 CHEMISTRY—WHAT CAN YOU SYNTHESIZE?

3.1 Chemistry of Spin Labels

The nitroxide group has been incorporated into a large number of starting materials and biologically important analogs, as illustrated in Figure 5. The compounds I through XX represent the most common starting materials, while XXIII through LVI are analogs of these basic nitroxides. The series LVI through LXXVII has been important in membrane studies, as illustrated by their lipid nature. Since the chemistry of nitroxides has been covered in great depth in a book on this subject by Rozantsev [30], we consider only a few outstanding features in this chapter [31, 9]. The most common method of synthesis of the nitroxide group in spin label syntheses has been the hydrogen peroxide oxidation of the corresponding di-t-alkylamine (I or II) in aqueous medium with phosphotungstate and/or sodium tungstate–EDTA as catalyst. A second, slightly more expensive procedure involves oxidation of the amine precursor in ether with m-chloroperbenzoic acid [32]. This latter procedure is absolutely necessary in the oxidation of oxazolidine precursors (III) to the corresponding nitroxide. In the course of a spin label synthesis, the precursor nitroxide moiety is usually oxidized early in the synthesis, since peroxide or peracid treatment is frequently too harsh for the biologically important functionality to be incorporated into the final spin label.

There are generally two useful groups of starting nitroxides employed in spin label compounds. The first group is IV through XX, derived from triacetoneamine, the amine precursor of 2,2,6,6-tetramethyl-4-piperidone-1-oxyl (IV). Although many of

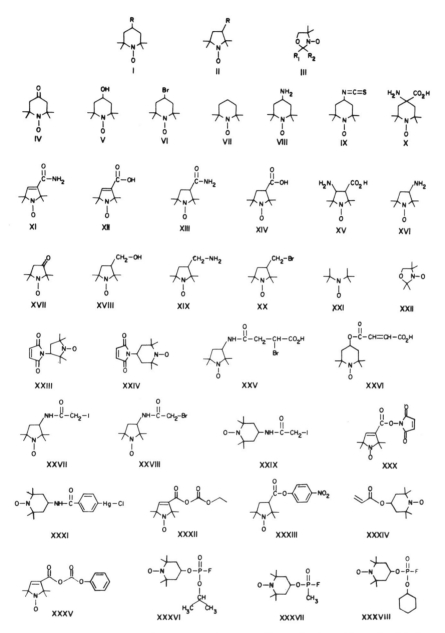

Figure 5 Structures of a representative sampling of nitroxide spin labels. References to the syntheses of these labels and those discussed in this chapter will be found in the Appendix. From reference 51 with permission.

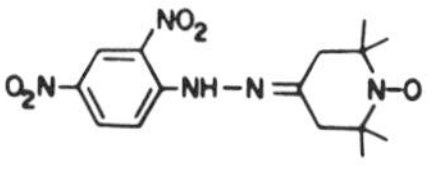

XXXIX 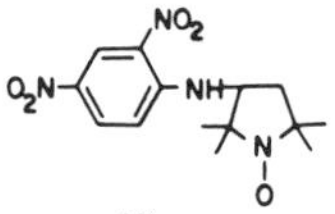   XL 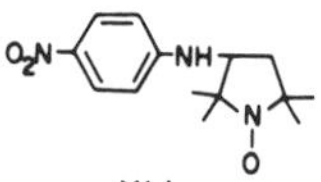   XLI 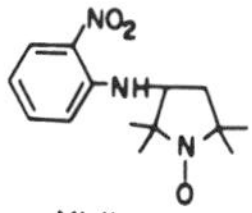   XLII

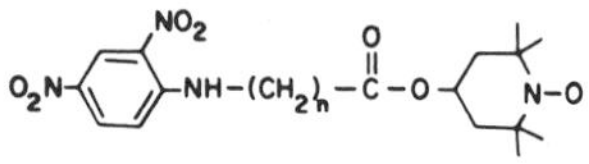

XLIII 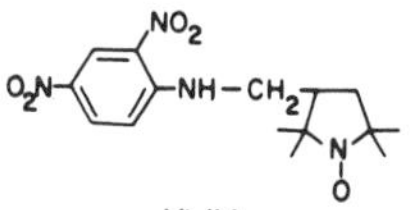   XLIV 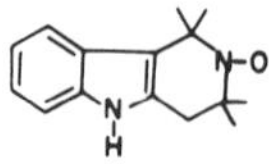   XLV

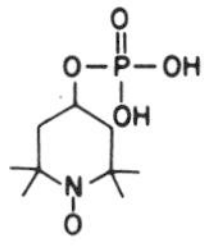

XLVI 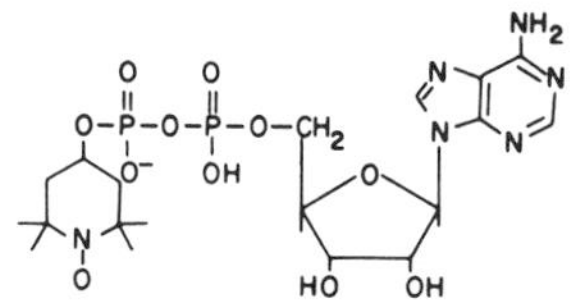   XLVII 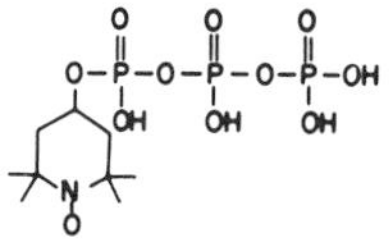   XLVIII

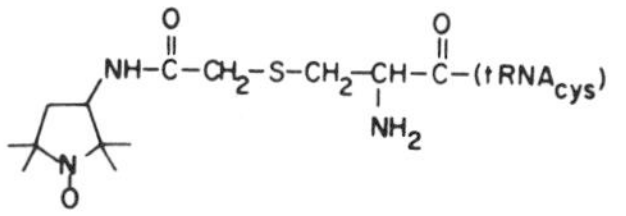

XLIX 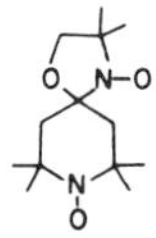   L 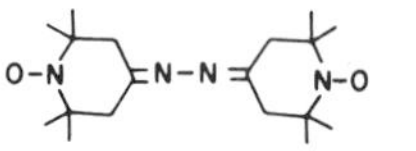   LI

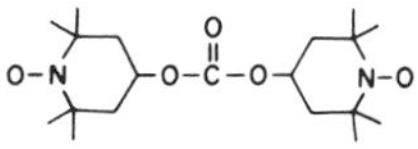

LII 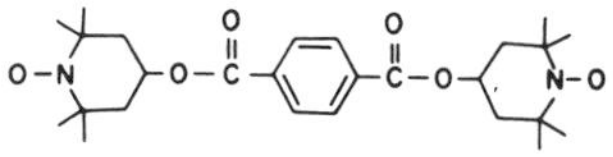   LIII

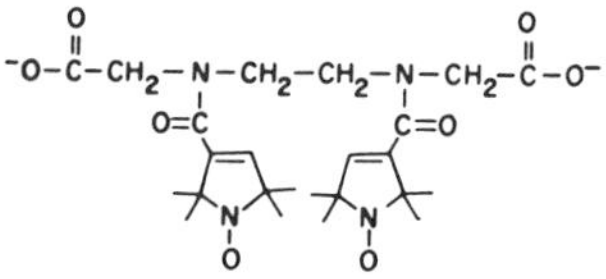

LIV 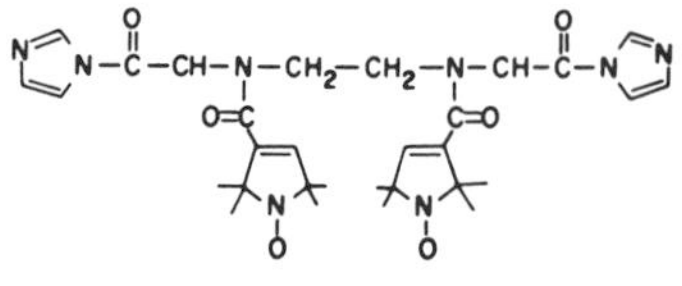   LV

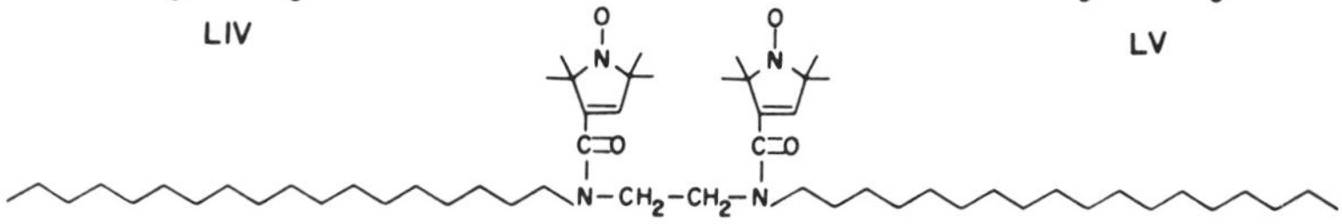

LVI

Figure 5   (*continued*)

14

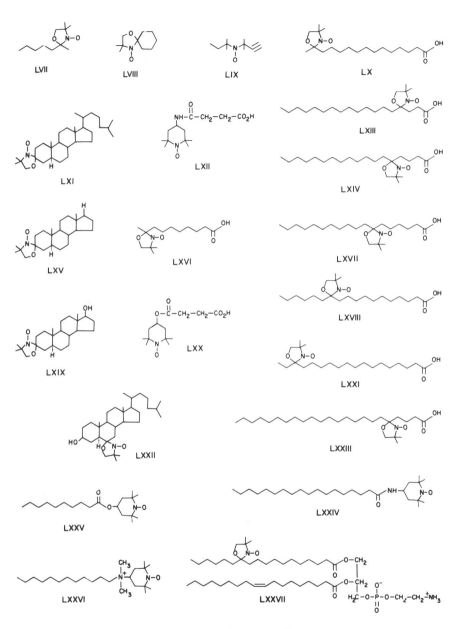

Figure 5 (*continued*)

15

these starting materials are now available commercially,* it is interesting to note that triacetoneamine is easily made from very inexpensive starting materials: acetone, $CaCl_2$, and ammonia [33]. The second, more useful, method of synthesis, which prospers from its wide general applicability, involves the synthesis of a protected 4,4'-dimethyl oxazolidine (III) analog

The most commonly used labels have been made from a ketone and 2-amino-2-methyl propanol (R_3 = R_4 = CH_3), although the only requirement is that the flanking di-t-alkyl groups R_1, R_2, R_3, R_4 be methyl groups or higher alkanyl groupings [32].

Given the basic starting materials above, the synthesis of diverse biological analogs is almost unlimited. The nitroxide moiety displays an unusual stability to most organic chemical treatments. It is stable in a pH range of 3–10, and is stable for shorter periods of time in strongly acidic solution. The nitroxide group is not reduced by KBH_4, nor in stoichiometric reductions of, for example, a carboxylic acid moiety with $LiAlH_4$ [34, 15]. Some mild or strong reducing agents, such as hydroxylamine or $LiAlH_4$ reduce the nitroxide to the N-hydroxylamine. However, it is also reversibly oxidized by exposure to air or the oxidizing agents discussed earlier [35]. In terms of biological stability it is quite stable except when exposed to ascorbate, glutathione, or accessible sulfhydryl groups [36, 37]. The last-mentioned case could present an undesired obstacle in protein studies, but if we consider that most experiments are designed to "localize" the spin label reporter group in a certain region of the protein structure, such side reactions may usually be negligible or absent. However, a sensitive measure of the accessibility of a protein's sulfhydryl groups as a comparison with present chemical methods could be accomplished through the employment of some nonspecific nitroxide, such as the piperidinol V. This measurement is made by following the rate of the concomitant decrease in esr signal accompanying reduction of the nitroxide by the sulfhydryl group [37].

*Some nitroxides are now sold by Aldrich, Eastman, Frinton, and SYVA.

The otherwise unusual stability of this class of nitroxides is due primarily to their inability to undergo disproportionation, a property inherent in almost all other *non-di-t*-alkyl nitroxides. Second, the

sterically protective α-methyl (or larger) groups prevent reaction with other nitroxides or bulky reagents. While one might assume that the lack of delocalization of the free electron in this formal three-electron bond would hinder stability, this group of nitroxides is more stable than many aromatic nitroxides.

Although it was mentioned that synthetic problems are at a very minimum for most spin labels, an unusual problem arises now and then. For instance, in the attempted preparations of benzoic acid esters and amides and *O*-glycosides of the six-membered piperidine I (V or VIII), or phosphatidic acid esters of the secondary alcohol pyrrolidine II (R = OH), synthetic difficulties arose either as a result of facile hydrolysis of the product [38–40], or also possibly side reactions to dehydrated elimination products [41, 42].

3.2 Molecular Structure of Nitroxide Molecules

The intimate binding of a spin label to a biomolecular site is of course dependent on the structure and geometry of the spin label. Several of the nitroxides in Figure 5 have been studied crystallographically. The potassium salt of the pyrrolinecarboxylic acid XII and its amide XI were both found to be completely planar structures [43, 44]. The N–O bond length was approximately 1.27 Å, which is intermediate between an N–O double-bond (1.20 Å) and single-bond length (1.44 Å) [45]. Surprisingly, however, the piperidinyl nitroxides, which exist in an overall chair conformation, were found to have the N–O bond slightly "out of plane" with the flanking trialkyl carbon atoms. In a study of the piperidinol V and subsequent refinement of the structure, it was found that the slightly longer (1.29Å), N–O bond was 16° out of plane [46, 47]. This was initially thought to create a problem in justifying the quite small isotropic electron-nuclear hyperfine coupling constant of 15–16 G observed for these nitroxides in solution, which was consistent with the assumption of a planar grouping of atoms. The slightly nonplanar structure, however, was rationalized theoretically not to be a significant distortion [47]. Lajzérowicz and co-workers have examined several of the piperidinyl nitroxide structures as well as others. They found the N–O group to be 30.5° out of plane in the radical, 9-azabicyclo[3,3,1]-nonan-3-one-9-oxyl [48],

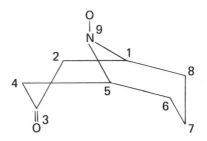

and 21.4° out of plane in the suberate [49].

A published report will be forthcoming on the structure of the (saturated) pyrrolidine derivative

XIV

in which there is a finite possibility of nonplanarity in the ring as well [50, 195]. The absolute structure of an oxazolidine nitroxide **III** is now known [196].

Instead of developing the syntheses of many of the biochemical analogs depicted in Figure 5, there will be remarks on some as they appear in their specific biochemical applications discussed later. (References to the syntheses and applications of all the spin labels described in this chapter, including Fig. 5, may be found in the Appendix.)

4 EXPERIMENTAL TECHNIQUE; SPECTRAL INTERPRETATION

The basic instrument requirement of a spin label experiment is an esr spectrometer, shown schematically in Figure 6. In brief, it consists of a source of electromagnetic energy, a microwave klystron, which is regulated in intensity by the fixed isolator and attenuator through a "magic tee" to the sample in the microwave cavity (see Fig. 6). The "magic tee" serves as a delicate balancing device (microwave bridge), which is designed so that the impedances of arms 2 and 4 are almost balanced, and also such that very little power reaches the detector (arm 3). The sample is in the microwave cavity, a "sample cell compartment" of critical dimensions. The sample cell is oriented carefully in a region of high microwave field density H_1 at the center of the cavity. The klystron frequency v is held constant in resonance experiments by an automatic frequency control circuit. Then, as the external magnetic field is scanned, the sample absorbs microwave energy when passing through magnetic field strengths H where the resonance condition is met. This microwave energy absorption at the sample is monitored as an imbalance in the microwave bridge by an increased power through arm 3 to the detector. The field modulation coils serve to implement the conversion of this sample absorption into a highly amplified first-derivative signal with a good signal-to-noise ratio (S/N).

It should be pointed out here that all these component parts of the esr spectrometer must be well understood by the prospective researcher before attempting to measure a spectrum. The spin label experiment requires that the operator be an esr spectroscopist as well as a competent biochemist. Although it is beyond the intended scope of this chapter, it must be emphasized that such variables as klystron (microwave) power, field modulation amplitude, magnetic field scanning rate, detector (signal filtering) time constant, and magnetic field inhomogeneity *all* may have artificial effects on the resultant spectrum if *all* are not properly adjusted. The reader is referred to the excellent article by Jost and Griffith for a more detailed discussion of these factors [51].

ESR SPECTROMETER

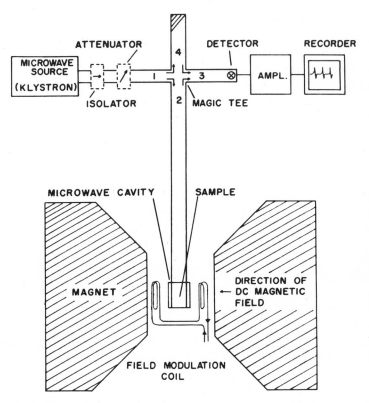

Figure 6   A block diagram of a simple esr spectrometer. From reference 51 with permission.

A common cavity geometry for experiments at X-band frequency is the rectangular type, such as that used in most Varian esr spectrometers. For high-sensitivity requirements in biological experiments, a thin, flat, rectangular quartz cell of dimensions approximately 6 cm X 1 cm X 0.3 mm inner internal diameter (minimum sample volume approximately 0.2 ml) is most commonly used. This specific geometry is necessary because of the high dielectric loss of water in aqueous samples. High-quality quartz is necessary because of its low content of trace paramagnetic metal ions and other radical centers. Such a configuration as described above allows one to detect spin concentrations below 10^{-6} M to an absolute limit of about 10^{-8} M. When operating through the 10^{-5} to 10^{-8} M range, the technique of multiple-scan time-averaging may become a necessity. In practice, however, if one can work at concentrations near 10^{-4} M, a clearly resolved low-noise resonance spectrum is usually obtained in a single scan. At these higher spin label concentrations, or under nonaqueous conditions, several different sample cell designs may be used.

For quick, fairly routine measurements where sensitivity is not a significant problem, we use an unbreakable Teflon sample holder, as shown in Figure 7. The sample cells are simply sealed 9-in. (length) disposable Pasteur pipets which are inserted into the quartz tube insert of the holder. The quartz insert serves as the replaceable part should the sample tube leak or break. The rigid fit of the Teflon piece to the Varian E-4531 rectangular cavity serves to orient reproducibly every sample tube at the cavity center. We have obtained a deviation of no more than 3% in signal peak height for a standard solution measured over several days in several (slightly varying width) Pasteur pipets.

Extreme caution should be taken when interpreting spectra, especially by a researcher new to this field. Besides adjustment of the instrument parameters discussed briefly earlier, there are several other possible chemical contributions to the resonance spectrum that may arise and which are related to the spin label–biomolecule complex under study: (1) other paramagnetic species (for instance, free, unbound spin label or paramagnetic transition metal ions), (2) solid, undissolved particles of spin label, and (3) spin-spin interactions resulting from high local concentrations of label in the biomolecular system. In special cases, such as a reversible equilibrium experiment, phenomenon (1) is to be expected; phenomenon (3) may also be significant under conditions in which particularly close sites on a biomolecule are labeled specifically.

Besides obtaining the spectrum of the spin-labeled macromolecule in solution, under suitable conditions we can frequently measure concentrations of bound and free species, or obtain kinetic data for a process involving a change in either the motional or magnetic state of the spin label. If the motional state of the nitroxide is changing significantly, the resonance lines depicting each state will be independently measurable. A change in magnetic state refers to a change in electron-electron interaction for spin labels initially at high "magnetic" concentrations ($>10^{-3}$ M), or to a case in which the nitroxide group is being selectively reduced by some chemical

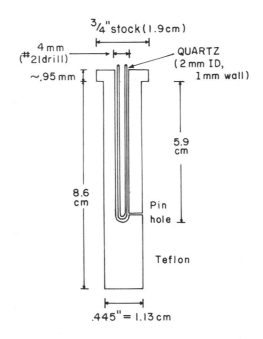

HOLDER and INSERT

Figure 7   Esr sample holder for routine measurements. The holder fits snugly into a Varian E-4531 rectangular cavity and accepts sealed 9-in.-long disposable Pasteur pipets as sample tubes. The top end of the disposable pipet is stoppered with a size-00 cork.

reagent such as described earlier. In studies of oriented biological samples, such as a spin-labeled protein crystal, the molecular orientation of the spin label may be determined [52]. Many of these techniques are aided by the use of an auxiliary small computer or analog-to-digital data collection device, although they are not absolute necessities in every case. We amplify on some of the above techniques in the specific applications section to follow. In addition there are several other general reviews on spin-labeling techniques now in the literature [10, 11, 51, 53, 54–57] including a complete text on both theory and applications [97].

5 APPLICATIONS TO SOME HYDROLYTIC ENZYMES AND RELATED EXAMPLES

5.1 α-Chymotrypsin

The first enzyme studied with spin labels was the well-characterized protease, α-chymotrypsin [58]. It was recognized even at this early stage in the development of the spin label technique that the problem of structural perturbations of the enzyme could be minimized by designing the spin label as a substrate analog. It had been shown by past workers that α-chymotrypsin catalyzed the hydrolysis not only of specific peptide linkages, but also of many "less specific" ester linkages [59]. In particular, p-nitrophenyl acetate, and other nitrophenyl esters had been shown to follow the kinetic scheme with α-chymotrypsin:

$$E + S \underset{}{\overset{K_S}{\rightleftharpoons}} E \cdot S \xrightarrow{k_2} E \cdot S' + P_1 \xrightarrow{k_3} E + P_2 \tag{7}$$

where E and S are the free enzyme and substrate, respectively; E·S, the reversible Michaelis complex; E·S', an acylated Ser-195 enzyme complex; P_1, the released nitrophenolate; and P_2, the subsequently released acyl substituent, which is a product of hydrolysis of the acyl–Ser-195 bond [59]. The complex E·S', was frequently isolable at low pH where the rate-limiting rate constant k_3 was quite small. In the spin-label experiments of Berliner and McConnell [58], the spin-labeled ester substrate,

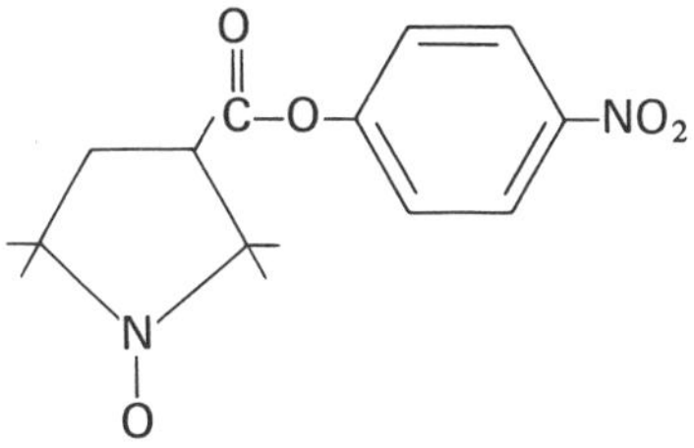

XXXIII

was employed as above, and a resultant esr spectrum was obtained which reflected the acyl enzyme complex (E·S') at pH 3.5. Its spectrum, shown in Figure 8a, indicated that the spin-labeled acyl

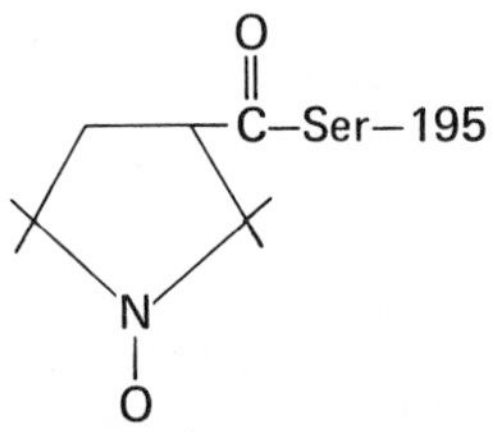

$$\text{O}$$
$$\|$$
$$\text{—C—Ser—195}$$

N
|
O
(E·S')

group was rigidly bound (strongly immobilized) at the enzyme active site. This was apparently caused by rather nonspecific "hydrophobic" binding forces, since the label did not resemble a typical amino acid substrate [59]. Furthermore, the conformational integrity of the covalent complex was demonstrated when the sample in Figure 8a was reversibly denatured at high urea concentration and the spectrum (Fig. 8a) returned after subsequent removal of the urea. Finally, Berliner and McConnell were able both to measure directly and confirm the kinetic step involving deacylation (k_3) by jumping

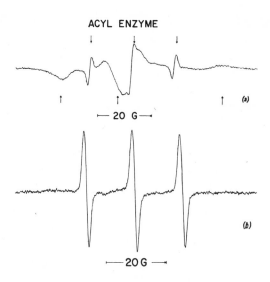

Figure 8 (a) Esr spectrum of the spin-labeled acyl enzyme at pH 3.5. Broad resonance lines, indicated by up arrows (↑) represent the acyl nitroxide moiety immobilized at the active site. The three narrow lines, indicated by down arrows (↓) arise from the "free" nitroxide carboxylate resulting from a very slow deacylation at this pH. (b) Resonance spectrum of the paramagnetic hydrolysis product, P_2 (nitroxide carboxylate) XIV. From reference 58 with permission.

the pH of the spin-labeled acyl enzyme to neutral pH and following the rate of appearance of the released, freely tumbling, carboxylate moiety (see Fig. 8b) [58].

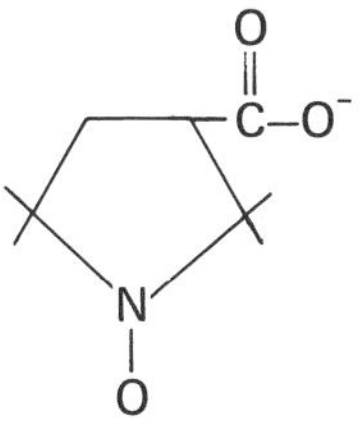

XIV

Here they obtained a *direct measure* of the rate constant k_3 rather than an apparent rate constant derived from the usual steady-state measurements with excess substrate [58].

Later, Kaiser, and co-workers examined the enantiomeric specificity of this substrate type by initially resolving the two enantiomeric forms of the carboxylic acid **XIV** and subsequently synthesizing the respective *p*-nitrophenyl esters [60]. Their kinetic data are summarized in Table 2. Remarkably, the specificity

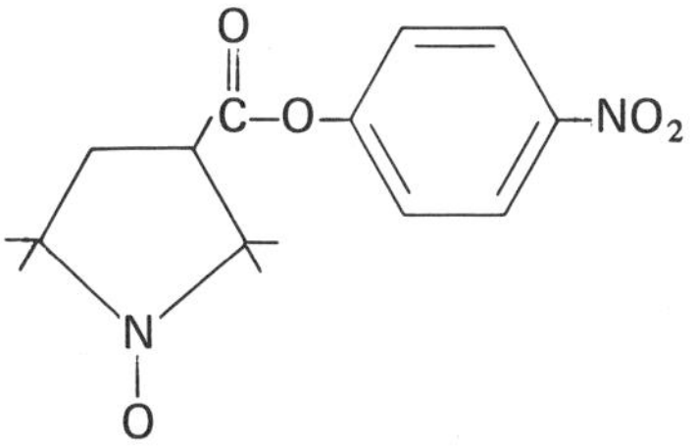

XXXIII

differences between the two enantiomers of this "nonspecific" substrate were not much greater than that found with a model enzyme system discussed later [61] (see Section 5.5). This was also reflected in the similarity between the respective acyl enzyme esr spectra for each enantiomer. Nevertheless, these differences should be sufficient enough to account for over 90% acylation of one enantiomer preferentially in a racemic mixture [62].

It was later shown in single-crystal studies that this strongly immobilized acyl group was indeed uniquely oriented at the active site of the enzyme [62]. The enzyme crystallizes as two symmetry-related dimers in the unit cell. The crystal unit cell contains four α-chymotrypsin molecules which may be considered

TABLE 2 KINETIC PARAMETERS OF BOTH ENANTIOMERS OF THE SPIN-LABELED NITROPHENYL ESTER SUBSTRATE XXXIII[a]

	(+)-Isomer	(−)-Isomer
K_S (enzyme-substrate equilibrium dissociation constant)	$(4.1 \pm 0.5) \times 10^{-4}$ M	$(5.1 \pm 1.2) \times 10^{-4}$ M
k_2 (acylation)	$(37 \pm 2) \times 10^{-2}$ sec^{-1}	$(4.1 \pm 0.4) \times 10^{-2}$ sec^{-1}
k_3 (deacylation)	$(5.2 \pm 0.2) \times 10^{-3}$ sec^{-1}	$(2.5 \pm 0.3) \times 10^{-4}$ sec^{-1}
k_3 (measured directly by esr at pH 6.9–7.3)	$(4.5 \pm 1.0) \times 10^{-3}$ sec^{-1}	$(2.3 \pm 0.2) \times 10^{-4}$ sec^{-1}
k_3 of unresolved isomers[b] Spectrophotometric		1.7×10^{-3} sec^{-1}
Esr		$(1.55 \pm 0.1) \times 10^{-3}$ sec^{-1}

[a] Conditions were pH 7.0, phosphate buffer, $\mu = 0.5$, 1% CH$_3$CN at 25°C, except where stated otherwise.

[b] Taken from the work of Berliner and McConnell at pH 6.8, ~ 25°C [58].

two dimers. The two dimers are symmetry-related by a twofold screw axis along the unit cell direction b. Furthermore, the two molecules *within* a dimer are related to one another by a molecular twofold rotation axis which is not part of the unit cell symmetry but is coincidentally parallel with the reciprocal lattice direction a* [63]. Thus in the esr experiment, if the magnetic field was in some general direction with respect to the crystal, no more than four sets of oriented spin-label spectra would be observed if the label were uniquely oriented in the enzyme. In these experiments, Berliner and McConnell found that when the magnetic field direction was perpendicular to b, only *two* single molecule spectra were observed [62]. The basis for this phenomenon is depicted in Figure 9 where the spin label is denoted by the *2p* orbital of the nitroxide nitrogen. At the special orientations where the applied field lies parallel or perpendicular to a twofold rotation axis, both N—O bonds subtend the same angle with respect to the external field, resulting in identical hyperfine and *g* values. Furthermore, in the very special orientation of the spin-labeled crystal where both the crystallographic symmetry axis b and the noncrystallographic "dimer" rotation axis a* were simultaneously perpendicular and parallel to the magnetic field, respectively, a *single* spectrum was observed representing a single molecular orientation [62]. Another *single* but different spectrum for the orientation where the magnetic field direction was perpendicular to the a* axis was observed and is shown in Figure 10. From these data the direction cosines were calculated for the vector perpendicular to the "planar" pyrrolidine ring. It was not very startling to realize that this spin-labeled acyl group did not and could not occupy the same site that the aromatic specificity group of a natural substrate does. This specificity pocket or slit, referred to as the tosyl hole, is too narrow to accommodate the "thick" pyrrolidine ring with its four bulky methyl groups which flank the nitroxide moiety [64]. Thus this unnatural acyl group is bound quite rigidly to an alternative binding area of the active site [65]. Since the absolute configuration of the (+)-enantiomer is now known to be *R*, a fairly unambiguous description of the binding of this spin label should be possible using the direction cosines above and the atomic model of the enzyme [50].

Later, Kosman, Hsia, and Piette [66] reported a study with some structural substrate analogs, based on the work of Cohen et al. with succinic, fumaric, and maleic acid esters and amides [67, 68]. Specifically, their summary of Cohen's model of the α-chymotrypsin active site was broken down into: (1) a hydrolytic site containing

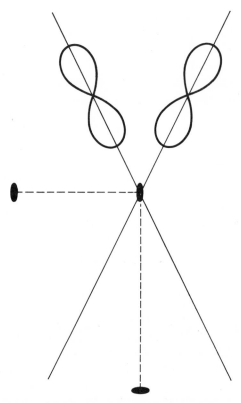

Figure 9   Two oriented spin labels related by a twofold rotation axis (vertical). The nitroxide groups are designated by the nitrogen $2p$ orbitals. Solid lines represent the unique (z) axis of each orbital. Dashed lines show the twofold axes resulting as a consequence of the inversion symmetry of the interaction of a paramagnet with the external magnetic field, that is, the g and hyperfine constants result from the angles subtended by the principal axes x, y, and z in Figure 2 with the applied magnetic field direction. The esr experiment cannot distinguish *up* from *down* in any orientation. From *Structural Chemistry and Molecular Biology*, A. Rich and N. Davidson, Eds., W. H. Freeman, copyright 1968, p. 146.

Ser-195; (2) an aromatic (*ar*) binding site (presumably the tosyl hole), which binds the aromatic side chain of a natural substrate; and (3) an amide-binding site (*am*) for the acyl amide group in a dipeptide or *N*-acylated substrate. Furthermore, they noted that Cohen's work suggested a transoid relationship between the hydrolytic site (Ser-195) and the aromatic (*ar*) site.

Their initial work involved the first three p-nitrophenyl ester substrates shown in Figure 11: a succinic amide (**LXXVIII**), a succinic ester (**LXXIX**), and a maleic ester (**LXXX**) [66]. The esr spectral results obtained by isolating at pH 3.0 the acyl enzyme intermediates formed from the reaction of each of these substrates with α-chymotrypsin were summarized as follows. (1) The succinic amide (**LXXVIII**) and ester (**LXXIX**) derivatives exhibited very similar spectra which reflected a moderately hindered spin label, whereas (2) the maleic ester derivative (**LXXX**) was characteristic of a covalently bound spin label with no additional noncovalent or "hydrophobic" interactions with the enzyme. They reasoned that the apparent lack of binding of the nitroxide group in **LXXX** was consistent with the cisoid configuration of the maleic group placing the nitroxide out of the active site. In an attempt to sort out these differences, they examined the effects of the strongly competitive

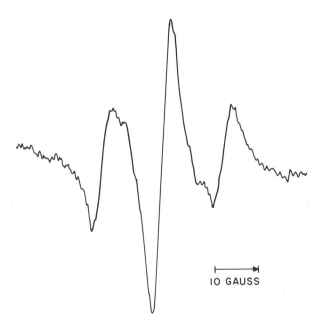

10 GAUSS

Figure 10 Esr spectrum of the acyl-chymotrypsin single crystal when the applied field is perpendicular to the crystallographic twofold screw axis b and also perpendicular to the noncrystallographic "dimer" rotation axis a*. A single spectrum is observed for this orientation, since all four spin label acyl groups are at identical orientations with respect to the applied field for this special case. A single, but different, spectrum is observed when the applied field is above but parallel to the a* axis.

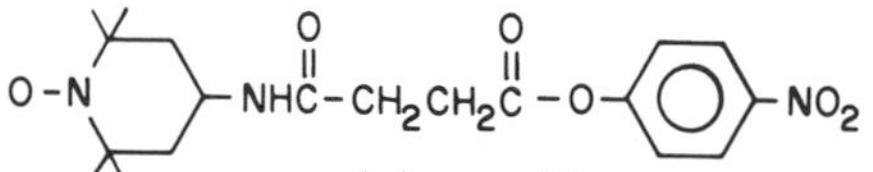

succinic amide
LXXVIII

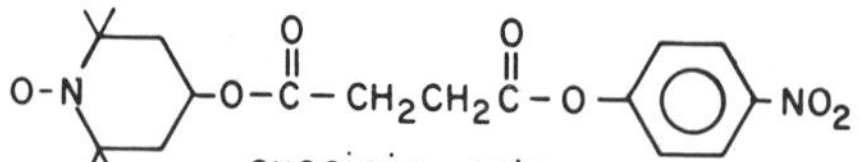

succinic ester
LXXIX

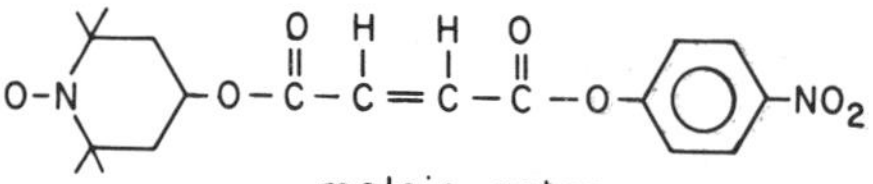

maleic ester
LXXX

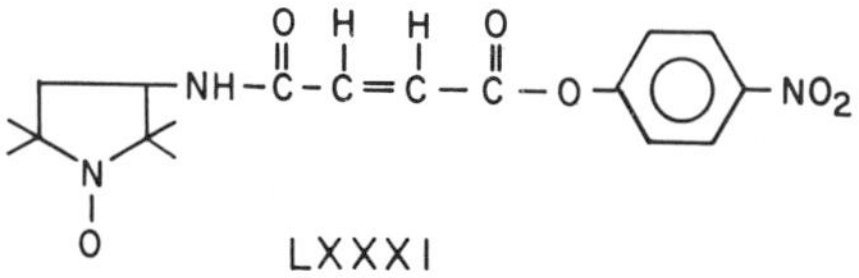

LXXXI

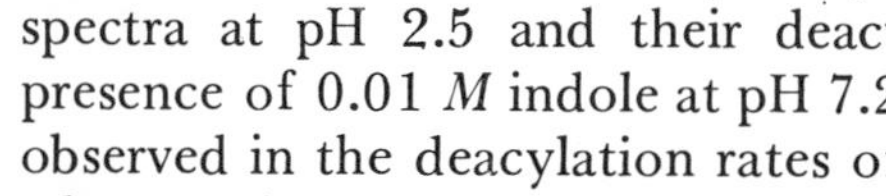

maleic amide
LXXXII

Figure 11 Some *p*-nitrophenyl ester substrates of spin-labeled succinamic, succinic, maleic, and maleamic acids. The terminology is adopted directly from Kosman et al. [66].

inhibitor indole on these acyl-enzymes, both by examining their esr spectra at pH 2.5 and their deacylation rates at pH 7.2. In the presence of 0.01 M indole at pH 7.2, a three- to fivefold increase was observed in the deacylation rates of both esters **LXXIX** and **LXXX**, whereas the succinic amide **LXXVIII** was retarded by a factor of approximately 3. Furthermore, the esr spectrum of the succinic amide (**LXXVIII**) derivative at pH 2.5 was shifted (reversibly) to a more strongly immobilized state in the presence of 0.01 M indole, while the corresponding ester (**LXXIX**) derivative was shifted (partially) to an extremely mobile state. Thus the more strongly immobilized succinic amide (**LXXVIII**) deacylated more slowly in the presence of indole, while the weakly bound succinic ester (**LXXIX**) deacylated more rapidly. With reference to the active-site model outlined above, they surmised that the succinic ester group

LXXIX was bound at the *ar* region, while hopefully the succinic amide LXXVIII was bound at the *am* site. The cisoid configuration of the maleic ester LXXX would preclude it from any specific binding. Since indole binds specifically to the crystallographic *ar* region [64], its effect on the ester LXXIX was consistent with the displacement of the spin label from this binding site. The model was tested further by examining these derivatives where Met-192 was oxidized by H_2O_2 to the corresponding sulfoxide [70]. The succinic ester LXXIX spectrum was affected in a fashion identical to that in indole, whereas the succinic amide was not. Since Met-192 had been implicated in the aromatic binding specificity of the enzyme [71, 72], its modification was consistent with indole no longer affecting *any* of the spectra or the deacylation rates to any significant extent. Furthermore, Kosman et al. suggested that the adverse "tightening" effect of indole on the succinic amide LXXVIII was a conformational change triggered at the *ar* site by indole. This, they surmised, was analogous to the binding of the aromatic moiety of a natural substrate which primed the enzyme for the acylation step via a productive structural change at the *am* site [66].

The above model of the active site was developed further in a later article. It had been shown meanwhile that the two maleic amides LXXXI [73] and LXXXII [74] displayed very hindered mobility as acyl-chymotrypsin derivatives. Specifically, the latter maleic amide derivative (LXXXII) gave a spectrum identical to the succinic ester LXXIX; it (LXXXII) was reversibly immobilized in the presence of indole as was the succinic amide (LXXVIII), and was also unaffected by Met-192 oxidation. Kosman and Piette noted that, while the model of Cohen indicated a trans relationship between the hydrolytic and aromatic site, there could exist a cis relationship between the hydrolytic and amide sites [74], and their results for the maleic amide (LXXXII) were consistent with this cis relationship. The more recent crystallographic evidence on the α-chymotrypsin structure indicates that nitroxides are too thick to fit in the narrow aromatic pocket [64]. This implies that the *ar* site for α-chymotrypsin substrates must include some additional part of the active-site structure besides the aromatic pocket. It may also be reasonable to speculate that the spectral changes observed upon exposure to indole could be explained simply on the basis of steric effects, a certainly less elegant model than that presented in the above work. It is interesting to note that no conformational changes were observed crystallographically in several chymotrypsin–small molecule complexes [64, 69]. A steric model for indole effects is presented in Section 5.3 in related studies with α-chymotrypsin.

In a subsequent article Kosman and Piette reported more extensive studies on the effects of active-site modifications [75]. In order to investigate more broadly the effects of oxidizing the critical methionine residue(s), they incorporated two diisopropyl fluoro-phosphate (DFP) analogs (see Fig. 13), a biradical label **LXXXIII**, and a monoradical label **LXXXVIII** which was identical to **LXXXV** except that a CH_3 replaced the C_2H_5. Oxidation of either Met-192 and/or Met-180 to the sulfoxide before reaction with the bulky biradical DFP analog **LXXXIII** yielded spectra indicating that only negligible changes had occurred in the environment of the spin label in the oxidized derivatives. However, when Met-192 or Met-180 was carboxymethylated, the biradical label (**LXXXIII**) was afforded greater motional freedom. The result was independent of the carboxymethyl moiety, its size, or its charge (see Fig. 12). The kinetics of inhibition of these alkylated chymotrypsins with either the biradical (**LXXXIII**) or the monoradical label was entirely consistent with steric effects rather than active-site conformational distortion. In these studies, the double label was kinetically quite sensitive to the size of the group on Met-192, whereas the smaller inhibitor was relatively unaffected. It was, nevertheless, suggested that an active-site distortion also occurred upon alkylation, since (1) the spin-labeled carboxymethylating reagent **XXIX** (the actual moiety on the enzyme is shown in Figure 12) exhibited a relatively mobile spectrum when bound to Met-192; (2) this moiety (**XXIX**) was only slightly affected by the subsequent introduction of a second quite bulky DFP analog (either dicyclohexyl fluorophos-phonate or the biradical **LXXXIII**); and (3) there was no evidence for spin-spin interactions between the three nitroxide groups in the experiment in (2). Presumably, the alkylated Met-192 side chain rotated "out" of the active-site region toward the solution environment. Unfortunately, they did not present data for the alkylated enzyme labeled with the smaller monoradical DFP label to reconfirm this conclusion.*

The spin label iodoacetamide **XXIX** at Met-192 was quite sensitive to pH-dependent conformational changes in the enzyme. A critical internal salt bridge between Ile-16 and Asp-194 appears to maintain the "active" conformation of the enzyme [77–79]. Titration of the Ile-16 α-ammonium group then should cause breakage of this salt bridge and subsequent destruction of the active-site conformation [78, 79]. The spin label mobility increased as the pH increased, with a sharp transition at approximately pH $\approx$ 9, the approximate pK_a of

*Preliminary studies elsewhere are in progress [76].

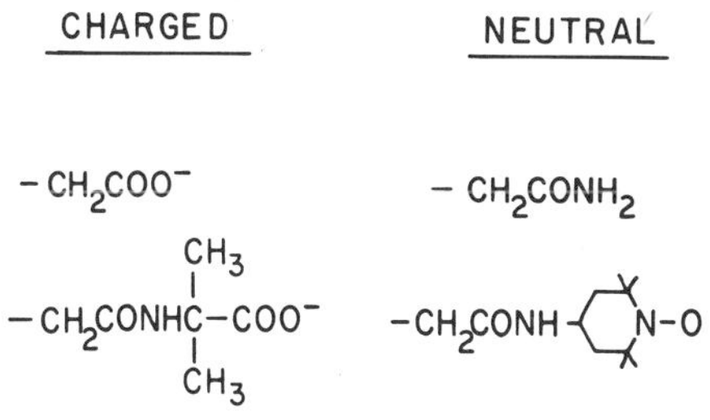

CHARGED   NEUTRAL

$-CH_2COO^-$ $-CH_2CONH_2$

$$-CH_2CONHC\!\!\begin{array}{c}CH_3\\|\\-COO^-\\|\\CH_3\end{array}$$

$-CH_2CONH-\langle\ \rangle\!\!-N-O$

Figure 12 Some neutral and charged alkylating agents specific for Met-192 of α-chymotrypsin. The modification yields an alkylated sulfonium (+)-ion.

Ile-16 [78]. Additionally, another conformationally sensitive pK_a was discovered at pH 3.8, presumably linked to the titration of some carboxyl group(s). In a later article Kosman demonstrated that the alkaline pK_a was shifted to higher pH in concentrations of indole sufficient to bind to the enzyme [80].

Met-192 of the inactive zymogen, chymotrypsinogen, was also modified with this spin label. The considerable rotational mobility observed, however, was shown to decrease simultaneously with the activation to α-chymotrypsin. The stereochemical mechanism of zymogen activation involves the movement of the Met-192 side chain out of the region comprising the specificity pocket to form the "lid" of the active site [81]. The change in mobility of the nitroxide during this process most certainly reflected this change, although one might naively guess initially that there should have been increased mobility as the (alkylated) Met-192 side chain moved out of the (restricted?) specificity site region. It is important to point out here that the determinants restricting the motion of a spin label can be both very complex and subtle—the more important factor to recognize is the sensitivity to changes in environment. In point of fact here, Kosman noted that the methionine side chain was reactive (exposed to solvent) in the zymogen only above 25°C, and therefore not the lowest thermodynamic state of the protein [80]. Of most significance was that chymotrypsinogen alkylated with XXIX was entirely different spectrally from the high-pH form of "active" chymotrypsin. It had been suggested in the literature that the breakage of the Ile-16–Asp-194 salt bridge at high pH was in concert with the enzyme shifting back to a "zymogen conformation" [82].

In an extension of this work, Kosman designed a His-57-specific alkylating agent LXXXIV analogous to N-tosyl-L-phenylalanyl chloromethyl ketone (TPCK) [80].

LXXXIV

This molecule, like TPCK, was assumed to alkylate specifically N-3 of His-57 [83]. It also alkylated anhydrochymotrypsin, a modified form of the enzyme in which the Ser-195 side chain is converted to a catalytically inactive dehydroalanine group [84]. Both spin-labeled derivatives displayed pH-dependent spectra which became more immobilized with increasing pH, in contrast to **XXIX**. Again a group of $pK_a \approx 8.9$ was implicated in the observed structural transition. Also, the binding of competitive inhibitors such as indole or methyl-N-(2-furylacryloyl)tryptophanate shifted this pK_a to higher pH, as in the case of the Met-192 derivative discussed earlier (pg. 33). The effect manifested itself by shifting the spin label to higher mobility. Kosman suggested that these results, together with those on the spin-labeled Met-192 were consistent with (1) the Met-192 side chain moving out of the specificity pocket upon zymogen activation, and (2) the further movement of Met-192 out into solution at high pH while Asp-194 moved out into the solvent at the mouth of the specificity pocket [80].

5.2 Organophosphate Spin Labels for Serine Esterases

Since organophosphates, particularly diisopropylfluorophosphate, had been generally useful as active-site-specific irreversible inhibitors for so many hydrolytic serine enzymes, it seemed only logical to design spin-labeled analogs. It follows, then, that a series of serine esterases could be tested for similarities in their active-site conformations where homologies existed in their primary structures. Almost simultaneously, Morrisett et al. [85] and Hsia et al. [86] reported the syntheses of several DFP-like spin labels (see Fig. 13). The former group synthesized **XXXVII** from methyl phosphonodifluoridate and the six-membered alcohol **V** [85, 87]. The latter

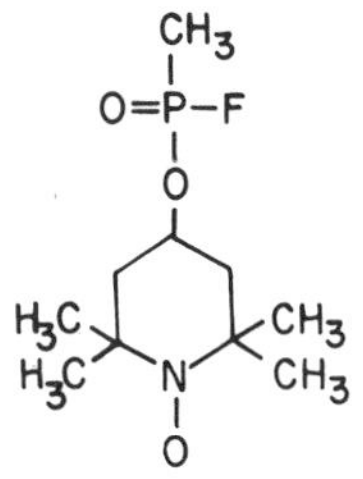

Figure 13 Some spin label DFP analogs. Labels **XXXVII** and **LXXXVII** are from references 85, 92, and 87. Labels **LXXXV** and **LXXXIII** are from references 86, and 88.

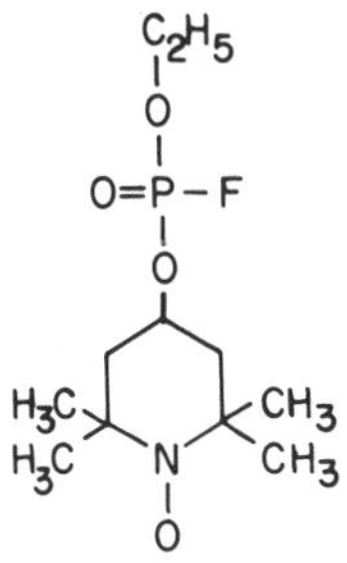

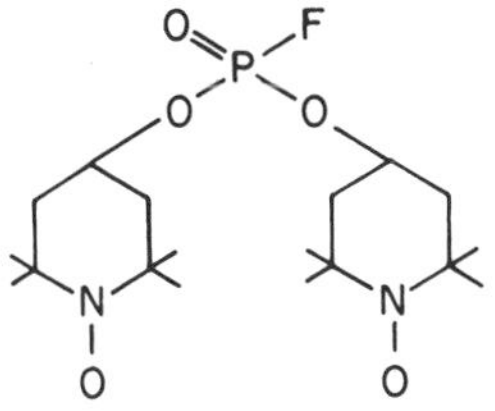

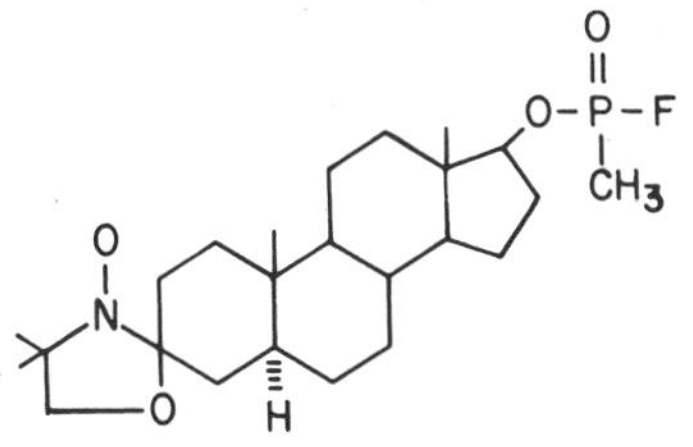

group prepared a series of compounds from phosphorus oxydichloro-fluoride and (sequentially) stoichiometric amounts of alcohols R_1OH

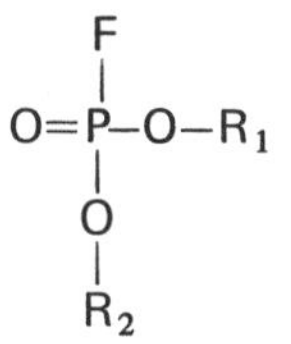

$$O=\overset{\overset{\displaystyle F}{|}}{\underset{\underset{\displaystyle R_2}{|}}{P}}-O-R_1$$

and R_2OH, where one or both alcohols were the six-membered piperidinol nitroxide **V** [86, 88]. The biradical DFP analog **LXXXIII** gives rather complex spectra whose complete interpretation is extremely difficult [89], although there is no doubt that the spectra are extremely sensitive to the environment of the label. The smaller monoradical labels **XXXVII** or **LXXXV** have been found to give identical spectral results when examined with the same enzymes [90, 91].

Three independent studies were reported which compared several serine esterases. Morrisett and Broomfield examined the enzymes acetylcholinesterase, trypsin, α-chymotrypsin, elastase, and sub-tilisin with the monoradical label **XXXVII** [92]. Hoff, Ooster-baan, and Deen used this same monoradical **XXXVII** in studies with atropinesterase, α-chymotrypsin, and subtilisin [93]. Hsia, Kosman, and Piette compared several cholinesterases, trypsin, α-chymotrypsin, elastase, subtilisins BPN′ and Carlsberg, and thrombin with the biradical spin label **LXXXIII** discussed earlier [88]. An abbreviated summary of these results is compiled in Table 3 where some of the major findings (and discrepancies) may be compared. The results are arranged in order of increasing mobility for each series of enzymes studied.

It is also important to note here that the extreme sensitivity of the spin label to local structural changes may create ambiguities in interpretation when small amounts of labeled enzyme impurities arise as background in the esr spectrum. For instance, in several reports in which trypsin was labeled at the active site, a moderate amount of the labeled enzyme was apparently autolyzed [88, 92]. This usually resulted in a two-component spectrum; that of labeled (intact) enzyme, and a narrow line component from autolyzed (labeled) enzyme. Unless the latter component is identified and removed, it can seriously confuse the interpretation of the spectra. (An example of this problem is shown in Fig. 14.) This misinterpretation was apparently made in the work of Hsia et al. with biradical-labeled (**LXXXIII**) trypsin [88]. They obtained a two-component (narrow line plus broad line) spectrum for trypsin labeled at pH 3.0, and a predominantly narrow line spectrum for trypsin labeled at pH 7.5. A rapid irreversible conversion of the pH 3.0 spectrum to the pH 7.5 spectrum was obtained when they raised the low-pH sample to pH 7.5. Their conclusions were that the enzyme assumed a low-pH conformation which was different and perhaps less stable than the pH 7.5 conformation, since the low-pH conformation underwent this irreversible conversion. However, the results are also consistent with a complete autolysis of the pH 3.0 derivative by some residual active trypsin in the sample. A similar but contrary discrepancy in high- and low-pH labeled trypsin also appeared in the work of Morrisett and Broomfield although they could not entirely rule out that autolysis was a possible alternative explanation for their results [92]. Berliner and Wong [91] carefully reinvestigated this problem with labels **XXXVII** and **LXXXV**, and showed conclusively that extensive enzyme autolysis did occur and did account for the above results (see Fig. 14). Since the autolyzed fragments were not completely removed by simple dialysis or

TABLE 3 COMPILED SUMMARY OF ESR SPECTRAL RESULTS ON A SERIES OF SERINE ESTERASES SPIN-LABELED WITH DFP ANALOGS[a]

	Morrisett and Broomfield [92][b]	Hoff, Oosterbaan, and Deen [93][c]	Hsia, Kosman, and Piette [88][d]
Reference:			
Spin label:	XXXVII	XXXVII	LXXXIII
High Mobility	Acetylcholinesterase (electric eel)		Fast exchange (high mobility) — Acetylcholinesterase (electric eel); Butyrylcholinesterase (horse serum); Acetylcholinesterase (bovine erythrocyte); Thrombin[g]; Elastase; Subtilisin (BPN')[h]
Moderate mobility	α-Chymotrypsin	α-Chymotrypsin	
Low mobility	Elastase; Trypsin[f]; Subtilisin (BPN')	Subtilisin (NOVO); Atropinesterase	Slow exchange (lower mobility) — Subtilisin (Carlsberg); Trypsin,[e] α-Chymotrypsin

[a] Each enzyme is positioned on a scale of increasing mobility. The third column with data from the biradical label **LXXXIII** is difficult to compare since any analysis beyond simply high or low mobility is difficult in interpreting these data.

[b] All spectra were reported at pH 3–3.5 except for the case of subtilisin (pH 4.3) at 25°C.

[c] All spectra were reported at pH 8.3 at 0°C, and therefore should be compared more cautiously with the other data. One problem is the higher viscosity of the solvent at this temperature. Another is "ageing" of the phosphonyl-nitroxide inhibitor on the enzyme [87, 92].

[d] All spectra were reported at pH 5.0. The temperature was not specified but is assumed to be 25°C.

[e] A "high-pH" form was probably a mixture of intact and autolyzed species. See [88, 91].

[f] A "low-pH" form was actually a mixture of intact and autolyzed species. See [92, 91].

[g] The authors noted that the high sequence homology between thrombin and trypsin was suggestive of conformational homology, whereas the spectra were quite different. It may be possible that their commercial thrombin preparation contained several impurities, as most reports on thrombin purifications point to extreme difficulty in removing the destructive antithrombin proteins [94]. This could account for the high mobility component in their spectra.

[h] With a monoradical label (**LXXXVIII**, the methyl analog of **LXXXV**), both Carlsberg and BPN' enzymes gave identical spectra and both contained in addition a sharp line component shown by Morrisett and Broomfield to be hydrolytically released nitroxyl phosphonate [92]. The difference in the two biradical (**LXXXIII**) spectra for BPN' and Carlsberg could simply be less complete removal of hydrolyzed nitroxide in the former over the latter.

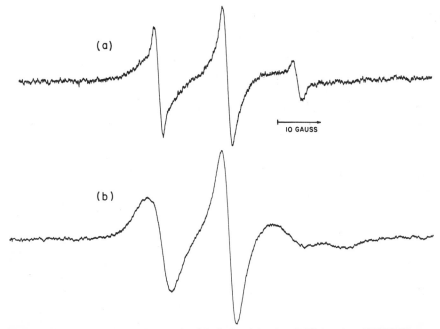

Figure 14 Esr spectra of trypsin labeled with the DFP analog **XXXVII** or **LXXXV**. Spectra were taken at X-band at pH 3.5, 0.02 M CaCl$_2$. (a) Spectrum of labeled trypsin containing slightly less than 1% (mole percent) of autolyzed labeled trypsin (narrow line components). (b) Spectrum of the pure α and β (intact) forms of labeled trypsin. From reference 91 with permission.

Sephadex G-25 chromatography, it is not difficult to understand how such a misinterpretation could occur.

It seems valid to speculate that a specifically labeled active site should interact to some degree with the nitroxide ring moiety, especially where the label displays saturation kinetics with the enzyme. Contrary to this hypothesis, almost all the results reported for organophosphate spin-labeled cholinesterases in Table 3 showed quite "free" unhindered spectra.* The probability of these highly mobile spectra being due to autolyzed forms did not seem to be as likely here. Acetylcholinesterase has been reported to be a quite stable enzyme [95], and all the spin-labeling results appear to be consistent with the notion that this is a representative conformation of the labeled enzyme (see Table 3).

---

*The spectrum reported by Morrisett and Broomfield, however, contained a small broad component as well [92].

In studying the results of Morrisett and Broomfield [92], we note that elastase, trypsin, and subtilisin gave similar spectral results, but of progressively increasing mobility in that order. The similarities in active-site conformation for these three enzymes were also consistent in the work of Hsia et al. [88]. Although Hsia et al. [88] reported a moderate amount of a highly mobile conformation with the subtilisins, this mobile component probably arose from hydrolytically released nitroxyl phosphonate, as found in the more detailed investigations of Morrisett and Broomfield with label **XXXVII** [92]. The exceptionally sensitive properties of the bulky biradical label were exhibited with elastase where there was a distinct increase in motional freedom relative to subtilisin and trypsin. However, Hsia et al.'s spectra of trypsin and α-chymotrypsin labeled with biradical **LXXXIII** were quite similar,* whereas the monoradical label **XXXVII** sensed real distinct differences between the two enzymes [92]. The results of Hoff et al. [93] also suggested that subtilisin (NOVO) was similar to α-chymotrypsin. However, since the latter workers reported their spectra at $0°C$, the results were undoubtedly all affected by the higher viscosity of these samples at this temperature.

As mentioned earlier, these studies were developed to investigate conformational homologies among several sequence-related enzymes; however, it is reasonable to speculate that results from just one spin label may not always be sufficient to yield a complete answer, especially where *no* spectral differences are found. A label with a moderate to high degree of motion may be less sensitive to the detailed structural features of its environment than a label that is more strongly immobilized. A comparison of the same group of enzymes with more than one label would certainly offer the strongest confirmation of the conclusions drawn in these studies. The next section describes just such an investigation.

5.3 A Conformational Comparison of α-Chymotrypsin and Trypsin Active-Site Structure

It was suggested above that a rigorous search or comparison of conformational homologies among related enzymes is accomplished

---

*The reported spectra of Hsia et al. were actually different [88]. However, if one mentally substracts that part of the trypsin spectra due (apparently) to autolyzed forms, the two spectra become more difficult to distinguish from one another.

by the incorporation of several spin labels, preferably of different geometry. The results of such a series of "isomorphous replacements" would be more conclusive than those obtained from just one label. It is especially desirable to probe a broad area of each active site. The spin label spectrum primarily reflects only mobility, not, unfortunately, a detailed picture of its binding. Thus the possibility of coincidental esr spectral features with two different enzymes for a particular label always exists. In attempting to circumvent such pitfalls, Berliner and Wong [96, 40] prepared several spin-labeled analogs of the structure where R = spin label.

$$SO_2F$$

LXXXVI

Aromatic sulfonyl fluorides had been shown to be satisfactory irreversible inhibitors for α-chymotrypsin, trypsin, and other serine proteases [98, 99]. Their approach was to prepare several ortho-, meta-, and para-substituted sulfonyl fluorides and compare their spectral behavior in α-chymotrypsin and trypsin. The spin labels are shown in Figure 15. There were some expected differences between the two enzymes, since several of the derivatives in Figure 15 were expected to bind in the chymotrypsin "tosyl" specificity pocket (assuming no added steric constraints). However, the trypsin active site may not allow a similar binding.* With para-substituted derivatives, however, it was expected that steric constraints would exclude binding in the chymotrypsin pocket, and therefore that any alternative nonspecific binding taken on by the label might be comparable to that in the homologous trypsin molecule. Specifically, published descriptions of the chymotrypsin structure, and structural constraints inferred from kinetic studies with aromatic substrates, dictated that bulky para substituent on the phenyl group should cause this moiety to be excluded from binding in the pocket [64, 81]. Aside from the difference in specificity binding pockets for

---

*Trypsin can bind the aromatic inhibitor, benzamidine, in its specificity pocket, although its binding overlaps that of diisopropylphosphoryl-trypsin (197). Unfortunately, the structure of tosyl trypsin is not known.

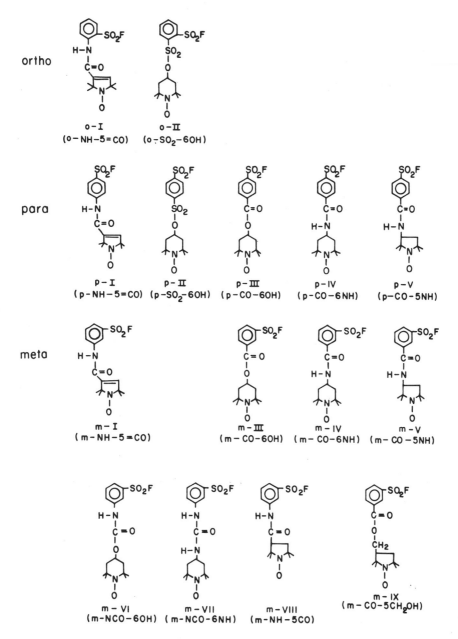

Figure 15 Structures of spin-labeled sulfonyl fluorides (**LXXXVI**).

each enzyme, their primary and tertiary structures are remarkably similar [100].

Some representative spectra are shown in Figure 16. It was obvious from comparing several of these derivatives that there were differences in the active-site conformations of each. However, since the chymotrypsin active site contains an aromatic specificity pocket which may bind the aromatic moieties of several of the labels in Figure 15, only those labels that *cannot* bind in this pocket could logically be compared with those on trypsin. Those derivatives considered, then, were all the para-substituted derivatives, p-I to p-V (Fig. 15), and also those meta derivatives shown by model building to be sterically prohibited from binding as well, m-I, m-III, m-IV. The remaining inhibitors then would bind in the tosyl hole and should reflect this unique feature of the α-chymotrypsin structure.

The alternate nonspecific binding region of α-chymotrypsin appeared nonetheless to be either sensitive to, or in structural proximity to, the tosyl hole. Dramatic effects were obtained when the esr spectrum was observed of each spin-labeled α-chymotrypsin derivative in the presence of saturated ($\sim$0.02 M) indole. Figure 17 shows some typical examples. This very effective competitive inhibitor for α-chymotrypsin had been shown from x-ray difference Fourier maps to bind specifically in the aromatic binding pocket [64]. Consequently, Berliner and Wong expected that those derivatives that bound in the tosyl hole would be "forced out" and displaced by an indole molecule. Yet, as a general rule, spin labels that were able to bind in the pocket (m-V, m-VI, m-VII, m-VIII, and m-IX) were least affected by indole (see Figure 17, bottom), while most of the other derivatives, which were sterically excluded from the pocket, were affected dramatically (see Figure 17, top). The striking feature in the spectra was the close similarity between chymotrypsin in indole and in trypsin (Figure 17, top). This similarity was also true for a number of labels (p-V, m-III, and m-IV), suggesting that indole displaced the spin label to a general alternate binding site common to both enzymes [96]. Apparently, the covalently bound derivatives in the tosyl hole were so tightly bound in this complex that their effective K_I values were orders of magnitude more favorable than the K_I of indole at this pH (the lowest pH value for which a dissociation constant is reported is pH 4.5, where K_I = 8 $\times$ 10^{-4} M [101]. However, those derivatives binding outside the pocket were either overlapping the entrance to the tosyl hole or were sensitive to enzyme conformational changes induced by indole binding. The former mechanism is actually the more appealing, and does not require the invocation of an induced fit or allosteric effect to

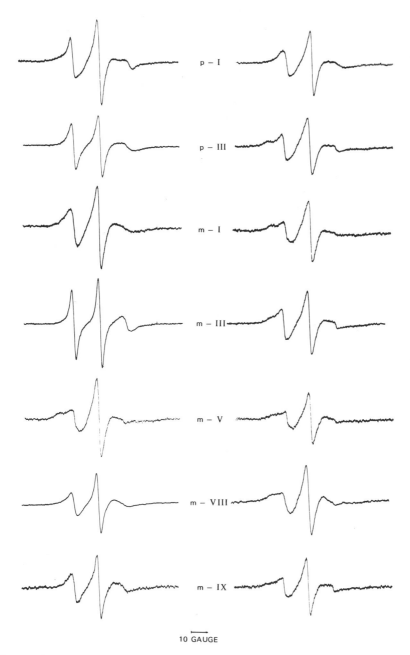

CHYMOTRYPSIN   TRYPSIN

p – I

p – III

m – I

m – III

m – V

m – VIII

m – IX

10 GAUGE

Figure 16 Some representative comparative esr spectra of α-chymotrypsin and trypsin spin-labeled with a series of sulfonyl fluorides. The spectra were necessarily measured at pH 3.5 to avoid desulfonylation.

43

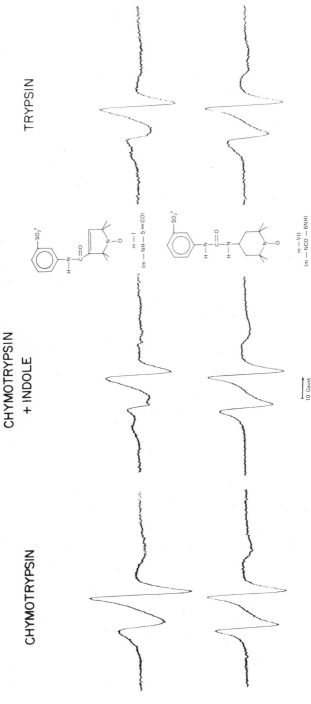

Figure 17 (Top) Labeling with sulfonyl fluoride spin label **m-I** (*m*-NH-5=CO). Esr spectra at pH 3.5 of α-chymotrypsin (left), α-chymotrypsin in saturated indole (center), and trypsin (right). This derivative binds outside the tosyl hole and is dramatically affected by indole binding. (Bottom) Labeling with **m-VII** (*m*-NCO-6NH). Esr spectra at pH 3.5 of α-chymotrypsin (left), α-chymotrypsin in saturated indole (center), and trypsin (right). This derivative can bind its aromatic group in the tosyl hole, therefore competing strongly with indole for this binding pocket.

rationalize the results. The change in spin label mobility upon binding of indole in the case of the pyrroline sulfanilide **m-I** was considered simply a shift in equilibrium from binding near the tosyl hole to the general binding region. No (detectable) conformational changes of any consequence were reported in the crystallographic studies of indole-chymotrypsin [64] and, second, the variety of spin label structures used in this work would not be expected to experience uniformly a "specific" immobilizing interaction allosterically triggered by the binding of indole. In total these results were strongly suggestive of the existence of a common general binding region for the two enzymes.

Finally, a seemingly obvious, yet commonly overlooked, pitfall with some spin labels became apparent during the studies of the five-membered derivatives **m-IX**, **m-VIII**, **m-I**, and **m-V** with α-chymotrypsin. An examination of the spectra from these four derivatives showed progressively increasing immobilization of the spin label in the above order. Although this trend was probably "real" in that it was a total reflection of the nitroxide's interaction with the immediate environment, it was also true from examination of space-filling models of the four spin labels that, because of the nature of the five-membered ring geometry, each had *intrinsic barriers* to rotation about the spin label phenyl linkage(s), which increased in almost exactly the same order as the immobilization observed in the enzyme studies.* Therefore, without additional evidence, the qualitative conclusions drawn from chymotrypsin derivatives of each of these four labels could have been completely ascribed to the labels themselves. A rigid binding of only the aromatic group to the active site in each case was sufficient to produce such results. In the case of the work cited above, however, the effects of intrinsic barriers to rotation and immobilization by the protein were distinguished from one another.

The general experimental approaches laid out in the studies discussed above are of extreme value as a tool for comparing conformations of structurally related proteins. The more subtle effects observed specifically with α-chymotrypsin did not create any problems or ambiguities in the comparative study, since the modus operandi was strictly to study *differences* between two or more enzymes.

---

*A CPK model of **m-VIII** appeared to have less rotational hindrance than **m-I**.

5.4 Protein Folding and Denaturation

An area of intense interest to many physical biochemists is the phenomenon of protein folding and denaturation. There is interest in both the thermodynamics and mechanism of these processes. Needless to say, the former objective is frequently more easily studied than the latter. Most physical studies in this field have involved (1) the measurement of enzyme activity, (2) the ultraviolet (UV), optical rotary dispersion (ORD), or circular dichroism (CD) changes of the protein as whole, (3) hydrodynamic methods, and (4) the observation of resolvable nmr line shape changes that (occasionally) may be assigned to specific nuclei. Aside from the nmr technique, the monitoring of specifically assigned conformational changes during protein unfolding has been difficult. Since a spin label at an enzyme active site senses the "active" intact conformation of this region of the enzyme, it should be capable of accurately reporting changes in this conformation during protein unfolding.

The unfolding of α-chymotrypsin in the presence of the denaturants guanidine hydrochloride and urea was studied almost concurrently by Morrisett and Broomfield [87] and Berliner [90], respectively. The studies in the two denaturants were easily compared, since both workers used the monoradical DFP spin labels XXXVII and LXXV which give identical esr spectral results (see Fig. 13). Morrisett and Broomfield also examined the pentacyclic steroidal DFP analog LXXXVII (see Fig. 13), and monitored all their derivatives by CD as well [87].

Three important questions were asked in these studies: (1) Does the protein unfold spontaneously and uniformly, or is the unfolding process a local, perhaps sequential, phenomenon? (2) Is the unfolding pH dependent within a nominal pH range? (3) Is there stabilization to unfolding in the presence of bound substrates or inhibitors and can this be correlated to, for example, the degree of immobilization of this bound group?

In the experiments in guanidine hydrochloride some interesting unexpected results were obtained. By monitoring the esr lineshape with increasing denaturant concentration, Morrisett and Broomfield [87] followed a transition to a highly mobile spin label spectrum as the active-site structure unfolded (or distorted). By esr the enzyme labeled with the small monocyclic XXXVII unfolded at the same denaturant concentration at either acidic (3.0) or neutral (6.6) pH. The bulky pentacyclic label LXXXVII, which gave a very strongly immobilized (tightly bound) spectrum, was also pH *independent* at the same pH values, but was considerably more stable to higher guanidine concentrations. However, upon following the CD change, $\theta_{230 \text{ nm}} - \theta_{222 \text{ nm}}$, which was suggested to monitor a perturbation of Trp-141, the opposite results were found. Both of the labeled enzymes showed pH-*dependent* unfolding behavior by CD in which pH 6.6 was more stabilizing than pH 3.0. Furthermore, by CD the enzyme labeled with the bulky pentacyclic label was somewhat less stable than either the monoradical-labeled enzyme, or native (unlabeled) enzyme, at either pH. It was obvious that the CD was monitoring a different and apparently quite independent change than were the two spin labels **XXXVII** or **LXXXVII** at the active site.

First, the intrinsic viscosity of the labeled enzyme underwent its sharp transition at some point intermediate to the CD and subsequent esr changes. Second, there was a pH dependence of this unfolding (by CD), but again it governed certain, but not all, of the structural changes occurring during the denaturation process. Third, in the esr study the strongly immobilized pentacyclic label (**LXXXVII**) served to stabilize preferentially the active-site conformation against denaturant, as compared to the smaller label. On the contrary, however, the CD results simultaneously showed that the large bulky label aided in destabilizing the local environment of the chromophore being monitored.

The esr results in urea turned out to correlate more closely with other spectroscopic techniques [90]. First, however, we must realize that urea and guanidine hydrochloride are two entirely different denaturing agents. Second, it quickly became obvious that the protein-unfolding phenomenon was dependent upon the region of the macromolecule monitored, and perhaps the particular monitoring technique as well. The monocyclic DFP spin-labeled enzyme **LXXXV** discussed earlier was investigated, as well as a serine-acylated chymotrypsin where the acyl group was:

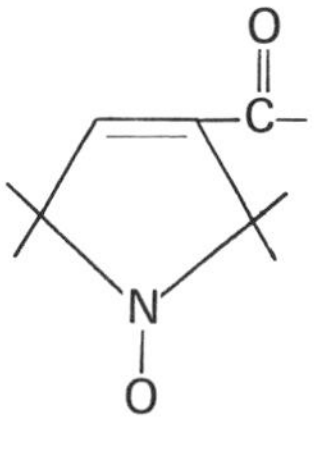

LXXXIX

(see Section 5.1 for a discussion of the acylation reaction). The latter acyl-chymotrypsin gave a strongly immobilized spectrum. Esr studies with the DFP nitroxide **LXXXV** at pH 2 and 5.5 showed that the (active-site) unfolding of the enzyme by urea was pH dependent, the stability to unfolding being enhanced at the higher pH. This correlated well with UV difference studies with DFP chymotrypsin reported in the literature [102]. Second, the tightly bound acyl chymotrypsin (**LXXXIX**) was found to unfold at exactly the same denaturant concentration as the less immobilized DFP analog **LXXXV**. Several other factors have been suggested to account for the unfolding results in these two studies. Assuming that there was an inherent pH stabilization of active-site conformation, it disappeared in guanidine hydrochloride, since this cationic denaturing agent would tend to "swamp out" or neutralize any electrostatic differences between the acidic and neutral conformation(s) of the enzyme. Second, the degree of immobilization of the spin label was not necessarily an indication of its stabilization of the overall enzyme structure.* A close examination of past literature indicated that molecules bound in the specificity pocket of the enzyme conferred more stability to it [103, 104]. In the studies above this was possibly the case with the large pentacyclic label, whereas the relatively small acyl group **LXXXIX** could interact with only a small region of the enzyme structure compared to the large pentacyclic label **LXXXVII**. Clearly, the degree of immobilization of a group bound at the active site of the enzyme was not the sole criterion for its stabilization of the protein's conformation.

5.5 Immobilized Enzymes and Enzyme Models

There is an increasing interest in bioorganic chemistry to (1) utilize advantageously enzymes in a routine manner for specifically catalyzed reactions in organic synthesis, and (2) to prepare synthetic organic polymeric catalysts having the basic properties of an enzyme. These endeavors are of strong industrial, as well as academic, importance.

5.5.1 Insolubilized enzymes. One successful approach to utilizing enzymes in a recyclable fashion has been accomplished by covalently

---

*There is no doubt that a group bound at the active site of this enzyme stabilizes its structure. A spin label acyl derivative (see Section 5.1), after heating at 100°C for 10 min, yielded a still strongly immobilized, bound spectrum (L. Berliner, unpublished results).

attaching the enzyme to an insoluble support [105]. When an enzyme is covalently attached to an insoluble polymeric support, however, it becomes virtually impossible to study any of its direct (dynamic) physical properties by normal spectroscopic methods. Kinetic analysis of the activity of an immobilized enzyme is complex, since substrate diffusion to the enzyme complicates the analysis of rate data. Usually such analyses are made with the insolubilized enzyme in a column or packed bed where the eluant is the substrate. Analysis of the product concentration and substrate residence time may be correlated with an observed rate and Michaelis-Menten constant [106]. An important structural question arises here which must be answered: Have any major conformational changes occurred in the enzyme structure as a result of this modification (immobilization)? Spin label investigations have been suggested as a critical approach to this question [107].

In the only complete study to date, Berliner, Miller, Uy, and Royer examined a spin-label phosphorylated (**XXXVII**, Fig. 13) trypsin derivative which was coupled to glass beads through aryl diazonium groups [107]. The initial goal was to verify that the active-site conformation was maintained after immobilization. In order to compare these two states, (spin-labeled) solution trypsin was examined in a saturated sucrose solution in order to approximate a macromolecular tumbling rate that approached immobilization (at least on the esr time scale). Immobilized trypsin was examined by two procedures. One was by subsequent labeling of preimmobilized (native) trypsin (labeled preimmobilized trypsin). The second was the immobilization (coupling to the beads) of prelabeled "solution" trypsin (prelabeled immobilized trypsin). The spectra in Figure 18 represent (a) spin-labeled solution trypsin, (b) labeled preimmobilized trypsin, (c) prelabeled immobilized trypsin, and (d) solution trypsin in saturated sucrose. The coincidence of the outer hyperfine extrema (separation approximately 53.5–54.5 G) in spectra (b) and (c) indicates that the nitroxide experienced the same degree of mobility in each case, and suggests that the active-site conformation was identical whether the enzyme was labeled before or after immobilization. The fixed enzyme retained considerable activity despite the immobilization process, and from the spin label experiments the active-site conformation was apparently unaltered by the presence or absence of a bound group at the active serine residue. The small narrow line components (arrows) in some of these spectra represent a very small percentage of autolyzed trypsin, which could not be completely removed from the solution or bead

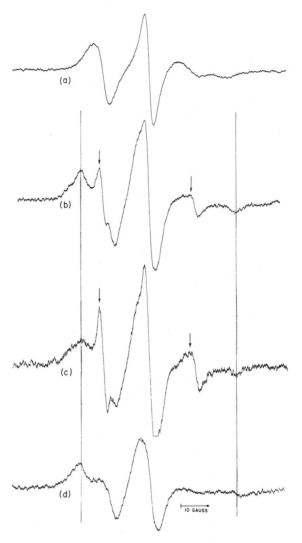

Figure 18 Esr spectra of solution and immobilized trypsins at pH 3.5 (0.05 M acetic acid, 0.02 M $CaCl_2$, 26°C, spin-labeled with the DFP analog **XXXVII** (Fig. 13). (a) Trypsin in aqueous solution (this is identical to the spectrum in Fig. 14b). (b) Labeled preimmobilized trypsin. (c) Prelabeled immobilized trypsin, and (d) Trypsin from spectrum (a) in saturated sucrose solution. The narrow line components which are particularly distinct in (b) and (c) represent a small fraction of autolyzed trypsin which could not be removed from these preparations (arrows ↓). From reference 107 with permission.

preparations. The spectra of solution trypsin in an aqueous (a) and a highly viscous medium (d) are indicative of the effects of macromolecular motion on the observed spin label spectrum. The mobility observed above for these labels was in the extremely sensitive intermediate tumbling range on the esr time scale, and therefore the sensitivity of this label (XXXVII) to changes in its immediate environment was extremely high. There were always obvious indications of the "intactness" of these immobilized enzymes from activity measurements, however, the subtle (structural) changes that occurred which were responsible for modified K_m or V_{max} parameters are more directly observable at the molecular level by reporter group methods. Although the spectrum of the solution enzyme in saturated sucrose (d) was similar to that of the immobilized enzyme, (b) and (c), the separation of the outer hyperfine extrema in (d) was approximately 1–2 G greater than in (b) and (c). This slight difference is attributable to several possibilities for which additional studies are contemplated: a subtle although not crucial change in active-site conformation due to the insolubilization reaction in (b) and (c); and the incomplete averaging of anisotropic nitroxide motion on the rigid immobilized enzyme matrix. The effects of high sucrose concentrations on active-site conformation seem less probable in view of the recent report by Timasheff [191]. Further refinement, such as a quantitative analysis of the specific surface residue(s) bonded to the matrix support and their distribution, is also in progress [108]. Since only one label had been examined and the differences between the various derivatives were small, it is fair to say that other labels should be examined to compare completely the similarity of these derivatives. Of course, the possibility exists that with larger spin labels, such as the fluorosulfonyl derivatives discussed earlier (see Section 5.3), it may be found that the active-site "periphery" is somewhat different in the solution and immobilized states.

The spin label on the immobilized enzyme was also a very convenient active-site titrant, just as a radiochemical label might serve. By either (1) doubly integrating the observed spin label spectrum, or (2) hydrolyzing the spin label off the immobilized enzyme to give a narrow line free spectrum, one may compare either spectral result from (1) or (2) with that of a standard. From the known weight of the enzyme beads and the percent of protein coupled to the beads, one may calculate both the active-site "molarity" and the percentage of active enzyme to total protein. For instance, in the study above with trypsin coupled to arylamine glass beads, the sample in Figure 18*b* indicated that 40% of the total protein was an active form of the enzyme. With this enzyme preparation, this number was somewhat misleading since the commercial trypsin starting material contains a substantial percentage of autolysis products. Therefore what additional enzyme activity was lost on account of the immobilization reaction was not significant.

Aside from just one report of a fluorescent reflectance measurement on insolubilized trypsin and α-chymotrypsin, no other "direct" approaches have been published [109]. The future for spin label studies of immobilized enzymes appears to be very bright, since the esr technique is one of the few physical methods to which the insoluble support is transparent. Also, the hyperfine parameters measured for labels on the *immobilized* enzyme reflect the infinite viscosity limit. Assuming that an immobilized enzyme is conformationally unaffected, this reference state is useful in calculations of spin label–protein rotational correlation times and protein rotational diffusion rates (see discussion in Section 2.1).

5.5.2 **Enzyme models.** The work so far in this area has yielded very promising and impressive results. Paton and Kaiser, in studies of a well-characterized cyclodextrin, the toroidal polysaccharide cyclo*hepta*-amylose, reported the direct physical observation of a noncovalent "Michaelis" complex in the model enzyme-substrate reaction [110]. Cyclo*hepta*-amylose had been shown by Bender and others to catalyze the hydrolysis of phosphate and carboxylic acid esters via intermediate formation (inclusion complexes) [111, 112]. The kinetic scheme for the hydrolysis of phenyl esters was demonstrated to be similar to that shown earlier for α-chymotrypsin and *p*-nitrophenyl acetate:

$$E + S \underset{}{\overset{K_S}{\rightleftharpoons}} E \cdot S \xrightarrow{k_2} E \cdot S' + P_1 \xrightarrow{k_3} E + P_2 \qquad (8)$$

where E is the cyclodextrin and K_S is the cyclodextrin-substrate dissociation constant. At pH 9.6 and 25°C the cyclodextrin catalyzed the hydrolysis of the nitrophenyl nitroxide ester

m–XXXIII.

with k_3 = 6.9 $\times$ 10^{-3} sec^{-1} and K_S = (7.5 ± 0.6) $\times$ 10^{-4} M. If the reaction was stopped by lowering the pH to 5.75, an acyl intermediate was isolable. However, if fresh cyclodextrin and ester were mixed at pH 5.75, where the studies reported with other esters suggested that a reversible noncovalent binding should occur [111, 112], the hydrolysis rate was quite small. The results are summarized in Table 4. Thus, at pH 5.75, they found that a substrate–cyclo*hepa*-amylose noncovalent complex (E·S) was observed with a nitroxide rotational tumbling rate (inverse of τ_c) slightly faster than that for the acyl cyclodextrin derivative (ES′). Furthermore, the K_S derived from an esr "equilibrium titration" measurement agreed well with that for the hydrolysis reaction at pH 9.6 [110]. It was extremely important that the spin-labeled ester be 100% pure, as any free hydrolyzed nitroxide carboxylic acid would give a spectrum superimposing that for the bound noncovalent intermediate. With any free paramagnetic impurities present, an apparently faster nitroxide tumbling rate can result.

In another study, Flohr, Paton, and Kaiser examined the enantiomeric specificity of an analogous cyclodextrin, cyclo*hexa*-amylose, in its catalyzed hydrolysis of the same nitroxide ester, m-XXXIII [61]. (There is an asymmetric carbon alpha to the carboxylic group in the nitroxide.) One might at first expect no specificity difference for either enantiomer in such a model enzyme system; however, the rate behavior of the hydrolysis of the racemic substrate showed a distinct (initial) fast reaction accounting for the hydrolysis of 50% of the substrate present, followed by a much slower hydrolysis of the remaining 50%. They resolved the starting acid, synthesized the respective m-nitrophenyl esters (+) and (−), and determined their kinetic parameters (see Table 5). Both the rate and dissociation constants for the fast reactant in the racemic mixture (±) were in good agreement with those obtained for the (+)-enantiomer

TABLE 4 ESR MOBILITY AND DISSOCIATION CONSTANTS FOR THE ASSOCIATION OF THE ESTER m-XXXIII WITH THE MODEL ENZYME CYCLOHEPTA-AMYLOSE

Species	Rotational Correlation Time of the Nitroxide, $\tau_c \times 10^{10}$ (sec)	Dissociation constant, K_S (M) and Method
Free label (m-XXXIII)	0.35	
Michaelis complex (E.S) (pH 5.75)	3.34	$6 \pm 2 \times 10^{-4}$ (Esr titration)
Spin-labeled acyl-cyclo*hepta*-amylose (E·S') (isolated at pH 5.75)	5.04	$7.5 \pm 0.6 \times 10^{-4}$ (spectrophotometric kinetics)

TABLE 5   KINETIC PARAMETERS FOR THE REACTION OF CYCLOHEXA-AMYLOSE WITH THE ENANTIOMERS OF THE SUBSTRATE m-XXXIII

	Racemic (±)-Ester		(+)-Ester
	Fast Reaction	Slow Reaction	
k_2 $(\text{sec}^{-1})^a$	0.022 ± 0.002	0.0032 ± 0.0006	0.025 ± 0.002
K_S $(M)^a$	0.013 ± 0.002	0.013 ± 0.005	0.019 ± 0.002
k_3 $(\text{sec}^{-1})^b$	1.1×10^{-4}	1.1×10^{-4}	1.1×10^{-4}

[a] pH 8.62 (tris-HCl), 25°C.
[b] pH 9.76 (carbonate), 25°C.

alone. Furthermore, it is of interest to note that the specificity difference for the two enantiomers showed up only in k_2, the rate constant governing the acylation step. These results were quite novel in two respects. The cyclo*hexa*-amylose specificity displayed here was the most marked reported to date, and ironically the homolog, cyclo*hepta*-amylose, displayed no such kinetic discrimination in its catalysis of the racemic substrate [110].

5.6 Lysozyme

Lysozyme has the distinction of being the first real enzyme to be described crystallographically at high resolution [113, 114]. This hydrolytic enzyme is specific for glycosidic linkages of β-$(1 \rightarrow 4)$-N-acetylglucosamine (NAG) oligomers and, in a bacterial cell wall, N-acetylmuramic acid (NAM)-β-$(1 \rightarrow 4)$-NAG oligomers. Furthermore, a crystallographically inferred mechanism required that the oligosaccharide substrate undergo steric distortion at the reducing end of the saccharide of the susceptible glycosidic linkage. In particular for a hexasaccharide substrate A-B-C-D-E-F, saccharide D must convert from a chair to a half-chair conformation. This distortion was also consistent with a proposed mechanism for the hydrolysis [115]. Here was a case in which it was hoped that appropriate application of spin-labeled analogs would shed more evidence on this mechanism. Particularly, one would imagine that a piperidinyl nitroxide, such as V, would serve as a good structural analog for a saccharide aglycone. Unfortunately, several attempts to synthesize NAG-V, glucosyl-V, or mannosyl-V glycosides failed

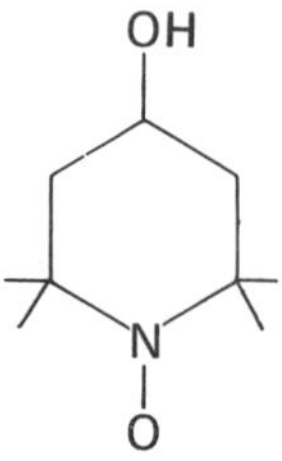

OH

(structure V)

V

dismally [38, 39]. It became quite clear that this particularly ideal nitroxide (V) was too labile to hydrolysis in these linkages for practical uses (see also the discussion in Section 3.1). Needless to say, the number of published spin label studies with this particular approach has been fewer than expected, although studies of other aspects of the protein's structure have yielded valuable results.

In an early study Berliner examined the interaction of the NAG analog XC with the enzyme [116].

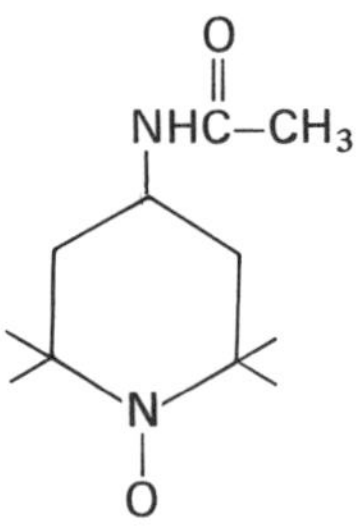

$$NHC-CH_3$$

(structure XC)

XC

The monosaccharide NAG had previously been shown to bind in the A and/or C binding sites of the enzyme, and to be a competitive inhibitor of glycoside hydrolysis [114, 117, 118]. The spin-labeled inhibitor analog XC also inhibited cell wall hydrolysis, but its poor inhibition association constant ($K_{assoc.} = 15M^{-1}$ at pH 7.0, 25°C) rendered any esr measurements practically impossible. A direct observation of the enzyme-inhibitor complex in solution was obscured by an enormously large free spectrum due to the unassociated spin label. However, the esr spectrum of polycrystalline lysozyme-inhibitor XC complex contained a resolvable strongly immobilized component. A more intriguing result was obtained from a crystallographic difference Fourier study of this enzyme-inhibitor complex [116]. At 6 Å resolution a new unexpected binding pocket was discovered which was more highly occupied than the expected binding at sites A or C. The spin label resided in a hydrophobic pocket near the surface of the molecule, which contained Trp-123 as a reference point. This pocket was far removed from the active site and was not known at the time to have any obvious function. As discussed below, this site was not a crystal artifact but a principal binding site for the acetamide XC in solution. Two cautionary notes were realized from these studies. First, an analog designed to interact with the active site of an enzyme may in fact bind preferentially elsewhere, and this should always be tested. Second, it is desirable to correlate both the crystallographic and the solution results in order to assess the total significance of a biochemical phenomenon, such as above. A model with this site is shown in Figure 19.

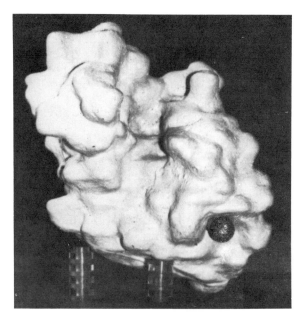

Figure 19   A space filling model of the lysozyme molecule. The dark ball is the spin-labeled inhibitor XC bound at the hydrophobic pocket near Trp-123. The active site is located in the vertical groove or "cleft" in the center of the molecule.

The above work was followed about a year later by an elegant structural study of the enzyme by paramagnetic nuclear relaxation techniques. In a "spin label induced nuclear relaxation study," Wien et al. quite convincingly "mapped" the lysozyme structure *in solution* [38]. The technique involved a careful measurement of the paramagnetic broadening of specific proton resonances caused by specifically incorporated nitroxides. The general methods have been employed quite extensively in enzyme systems containing paramagnetic metal ions by Mildvan and Cohn [119], and in a few examples with nitroxides by Mildvan and Weiner [120], Krugh [121], Bennick et al. [122], Roberts et al. [123], and Kornberg and McConnell [124]. The principal paramagnetic relaxation phenomenon in the spin label case is usually electron-nuclear dipolar coupling ("through space coupling"), since the electron and nucleus of interest are usually not close enough to permit isotropic (contact) coupling (as in ligand formation).

The conditions of the lysozyme experiments allowed the use of a simplified relationship for the dipolar-induced transverse relaxation rate $1/T_{2m}$ as follows

$$\frac{1}{T_{2m}} = \frac{S(S+1)\gamma_I^2 g^2 \beta^2}{15 r^6} \left[4\tau_c + \frac{3\tau_c}{1 + \omega_I^2 \tau_c^2} \right] \qquad (9)$$

where γ_I is the nuclear magnetogyric ratio, r is the electron-nuclear separation, ω_I is the nuclear Larmor precession frequency, τ_c is a complex correlation time for the dipolar interaction, and S, g, and β are defined as in Section 2.1. Equation 8 holds if ω_S, the electron Larmor precession frequency, is much greater than ω_I, and τ_c is large enough such that $\omega_S^2 \tau_c^2 \gg 1$. The dipolar correlation time is actually a complex function of several correlation times involving at the least the rotational correlation time of the vector joining the electron and the nucleus, the electron spin relaxation time, and the residence time of the nucleus-protein complex if the nucleus is on a molecule which is rapidly exchanging at some protein binding site. Assuming that the correlation time τ_c can be calculated or approximated, the electron-nuclear distance can be expressed as a direct function of the paramagnetic relaxation rate in the protein

$$r \propto (T_{2m})^{1/6} \qquad (10)$$

since the electron magnetic moment is about 700 times that of, for example, a proton, the paramagnetic dipolar interaction is frequently the dominant relaxation mechanism in these systems. The greater obstacle in studies of this type is in evaluating τ_c.

Wien et al. employed the spin labels shown in Figure 20 for their studies. The two saccharide analogs **XCI** and **XCII** were easily synthesized from the primary alcohol nitroxide **XVIII**. These were expected to bind to the enzyme in a similar fashion to those of

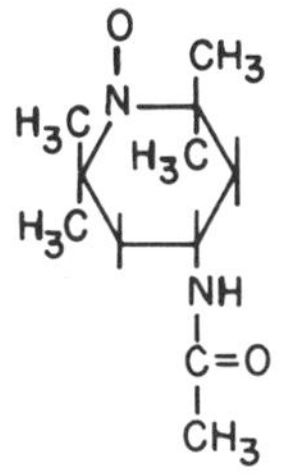

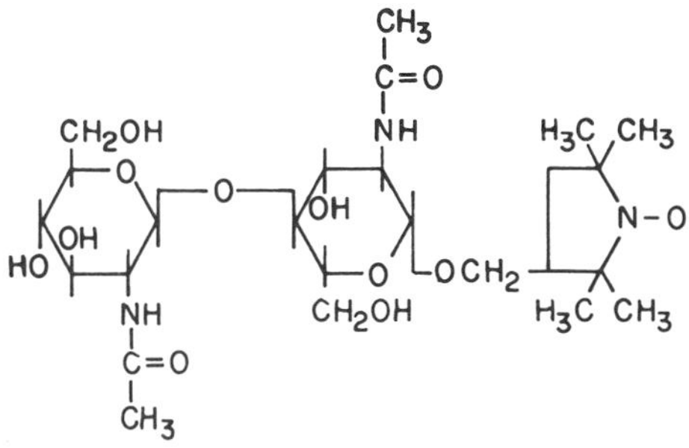

Figure 20   Spin label reagents which bind to lysozyme. All these nitroxides were employed in the nuclear relaxation studies of Wien, Morrisett, and McConnell [38].

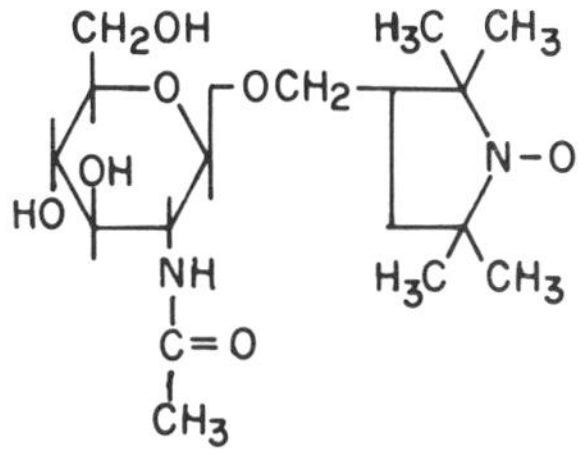

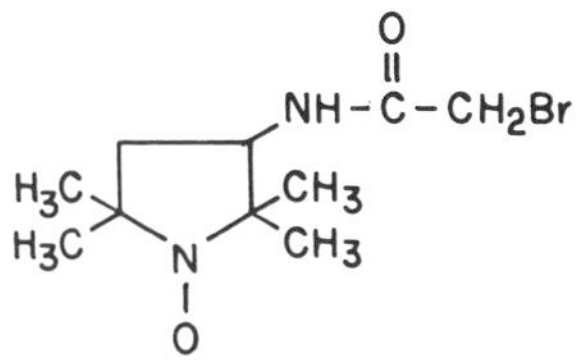

known NAG oligomer analogs. The piperidinyl acetamide **XC** was known from the work cited above to occupy preferentially the hydrophobic pocket near Trp-123 [116]. Finally, the bromo-acetamide analog **XXVIII** was shown to alkylate specifically the only histidine (residue 15) in the structure [38]. This histidine in its native state also has a resolvable C-2 imidazole proton resonance at 100 MHz.

A complete series of triangulation experiments was then carried out as shown diagrammatically in Figure 21. The spin label-induced relaxation was measured of:

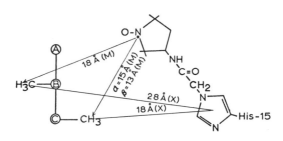

MEASUREMENTS USING
NON-COVALENT SPIN LABELS

MEASUREMENTS USING
COVALENTLY SPIN-LABELED LYSOZYME

Figure 21 A schematic diagram comparing distances between groups on lysozyme as determined by measurements from an x-ray derived model (X) and from paramagnetic nuclear relaxation studies (M). The lines designate the electron-nuclear interaction examined in each experiment. See the text for a more detailed description. From reference 38 with permission.

1. The C-2 proton of His-15 by either the spin-labeled inhibitor **XC** near Trp-123, or by the spin label saccharide analogs **XCI** or **XCII**, believed to bind with the nitroxide moiety at saccharide subsite D of the active site.

2. The N-acetylamido methyl protons of di-NAG, which binds at subsites B and C, or of NAG which binds at site C, by spin label inhibitor **XC** near Trp-123 or by the **XXVIII**-alkylated N-3 imidazole of His-15.

The distances in Figure 21 are designated by an X if crystallographically derived, or by an M if derived from nuclear relaxation experiments. The agreements were quite good, and certainly displayed the powerful sensitivity of this spin label–nmr combination.*

5.7 Ribonuclease

Ribonuclease was the second enzyme to be completely described in terms of a three-dimensional crystallographic analysis [126]. It was also well studied by other chemical and spectroscopic methods, as well as having a completely known primary structure [127, 128]. This endonuclease, which specifically catalyzes the hydrolysis of the 3'-O–P bond of pyrimidine ribonucleotides, has been shown to have the following three amino acid residues implicated in its action: His-12, His-119, and Lys-41 [129–132]. Furthermore, these residues can be alkylated with α-halo carbonyl compounds such as bromoacetic acid. Heinrikson et al. showed that ribonuclease A is inactivated by bromoacetic acid at pH 5.5, concurrently with alkylation (carboxymethylation) at N-1 of His-119, and about 12% of the time at N-3 of His-12 [131]. Carboxymethylation of one histidine precludes alkylation of the other on *the same enzyme molecule*. At a higher pH (8.5), Lys-41 is predominantly modified instead [130]. In an attempt to examine the active-site conformation of this enzyme, Smith studied carboxyalkylation of the enzyme with the spin-labeled bromoacid analogs **XXV** and **XXVIII**, as well as the maleimide analogs **XXIII** and **XXIV** [133].

XXV   **XXVIII**

*Sternlicht and Wheeler had actually reported earlier preliminary studies of the paramagnetic broadening of lysozyme protons. The enzyme was labeled with the nitroxide maleimide **XXIII**; however, the specificity of the labeling was unknown. Furthermore, the paramagnetic broadening observed was nonspecific [125].

XXIII   XXIV

The bromoacetamide analog **XXVIII** inactivated the enzyme as expected, presumably at His-12 since a carboxymethylated N-3 imidazole histidine was isolated, while the maleimide derivative **XXIV** reacted with Lys-41. Spectrophotometric studies indicated that the overall conformation of the labeled ribonuclease was not significantly different from the native enzyme. In studies of the structural stability of the enzyme to perturbations of pH, substrates, inhibitors, phosphate, and temperature, Smith discovered a thermally induced structural change in the ambient temperature range 5–45°C, which had never been detected before by optical methods. With either label **XXVIII** (bromoacetamide) or **XXIV** (maleimide), a linear Arrhenius-type plot of rotational correlation time versus inverse temperature was obtained with a transition point at 33°C. This new transition was different from the optical "melting" transition around 61°C [134]. The duplicity of these results, with the bromoacetamide label on His-12 (**XXVIII**) in one case and on the maleimide Lys-41 (**XXIV**) in the other, suggested that both labels were monitoring the same structural region and that the changes measured did not depend strongly on the nature of the spin label as well. It was further pointed out that, since the principal "marker" for the optically detected structural transitions were three abnormal tyrosine derivatives [135], these tyrosines were certainly insensitive to the very subtle change observed above.

Another interesting feature of these spin-labeled enzyme derivatives was their apparent ability to bind RNA and synthetic polynucleotides. Upon binding of RNA to ribonuclease labeled with bromoacetamide **XXVIII**, the spectrum broadened, indicating a further restriction of motion of the label at His-12. Smith correctly pointed out, however, that this decrease in motion could be due to either (1) a specific complex whose binding affects the immediate environment of the spin label at the active site, or (2) a random nonspecific complex which intrinsically has a longer rotational correlation time than an isolated ribonuclease molecule. The latter case was ruled out, however, when it was observed that, for derivatives that were spin-labeled at lysine residues on the protein exterior, the relatively free motional state of the label was unaffected by the formation of the nucleic acid–enzyme complex. The binding of the synthetic polynucleotides poly-A, poly-G, poly-I, and poly-C produced effects similar to RNA on this spin-labeled derivative, whereas poly-U, a substrate for the enzyme, showed no effect on the spin label spectrum of this derivative. It was also surprising that poly-G produced an effect similar to those of the other polynucleotides. Although the enzyme had not been previously shown to hydrolyze poly-G, this did not preclude its reversible binding in some (specific?) manner. Second, poly-U was the only polynucleotide that did not have an intrinsically highly ordered structure at the pH of these experiments. Presumably, the secondary structure of the poly-nucleotide may have some influence on both the spin label immobilization, as well as the specific nature of the binding to the enzyme. Neither 2′-cytidylic acid nor RNA core (the oligonucleotide digest after a 24-hr treatment of RNA with active ribonuclease) produced any change in the spectrum as well. The results suggested that the binding of these smaller nucleotides may be somewhat different from that of the natural substrate RNA or other poly-nucleotides.

Similar experiments were carried out with ribonuclease S, an active noncovalent complex of the two peptides produced by subtilisin cleavage of the peptide bond joining residues 20 and 21 in the native A enzyme [136]. The larger fragment is called the S protein, while the 20-residue polypeptide is called the S peptide. The results obtained from labeling ribonuclease S with bromoacetamide XXVIII, for example, indicated that the spin label was slightly more mobile than in the intact A enzyme, but not drastically different. In RNA binding experiments it was shown that the spin-labeled enzyme responded in a manner similar to the intact A protein. It was concluded that the active sites of ribonuclease A and S were much more similar than expected on the basis of earlier work from other methods [133].

In later reinvestigations of the labeling of ribonuclease with label XXVIII, it was shown that by varying the conditions of labeling (such as label concentration and pH) a variety of spin label products could be obtained. For example, derivatives labeled at Met-29 (or -30), His-119, His-48, and His-105 have been completely characterized, and their esr spectra examined under a variety of conditions [137, 138]. The results summarized earlier are in most part corroborated by the more detailed characterization studies.

5.8 Carbonic Anhydrase

Bovine erythrocyte carbonic anhydrase is a zinc-containing metalloenzyme which catalyzes, reversibly, the hydration of CO_2 (EC 4.2.1.1 carbonate hydrolyase). It is specifically inhibited by aromatic sulfonamides, $ArSO_2NH_2$, where Ar may be homo- or heterocyclic. Chignell and co-workers reported a novel study with spin-labeled sulfonamide derivatives of the general structure

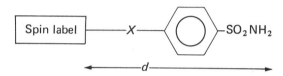

where X is essentially a methylene chain bridge of varying length [139, 140]. Such a "molecular dipstick" technique was used to measure the depth of the sulfonamide binding cavity of this enzyme. The basic technique was originally demonstrated with antibodies by Hsia and Piette [15, 141]. This application to carbonic anhydrase was especially exciting, since the complete crystal structure had been reported [142, 143].

Some of the spin labels and their dimensions are shown in Figure 22. In their initial work with the B isozyme of bovine carbonic anhydrase, Chignell et al. [139] found that as the mobility of bound sulfonamides was followed with increasing (fully extended) length, a transition to a much higher degree of mobility was observed at label **XCVIII**. They used the fully extended length of this label, 14.5 Å, as a measure of the depth of the sulfonamide combining site in this enzyme. In a further more detailed study of human erythrocyte carbonic anhydrase, the B isozyme was both reexamined and compared with similar studies on the C isozyme of the same species [140]. The results with the B isozyme, when further analyzed, pointed out some additional features of the enzyme active site that influenced the spin label spectra. For example, in going from spin label **XCVI** to **XCVII**, the nitroxide mobility in fact decreased somewhat. It was suggested that either the active site became *narrower* at this point and/or that this region of the cavity was more hydrophobic and therefore more strongly attractive toward the

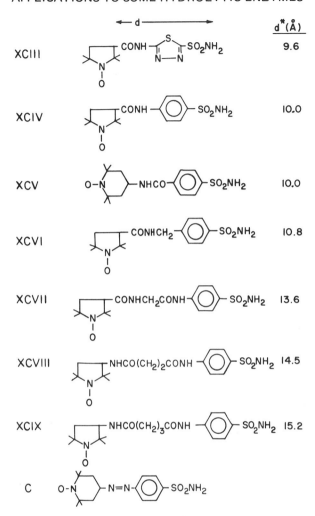

Figure 22 Spin-labeled sulfonamides for carbonic anhydrase. The distances d, where shown, were calculated for the fully extended conformation of the molecule from CPK models. The distances were taken from reference 139.

pyrrolidine ring. The next label **XCVIII** showed little interaction, as in the earlier work [139]. They suggested that the 14.5-Å dimension of this label should be regarded as an upper limit rather than the most probable depth of the cavity.

When the same observations were made with carbonic anhydrase C, a smooth transition was observed from sulfonamide **XCIV** to **XCVI** to **XCVII** *without* the intermediate degree of stronger binding as observed with label **XCVII** in the B isozyme [140]. The more confident estimate of the active-site depth here was 14 Å (the dimensions of label **XCVII**), which was in good agreement with the data from x-ray diffraction experiments (15 Å) on the C isozyme [142]. Remarkably, the sulfonamide **XCIX** displayed more immobilization with the C isozyme than did label **XCVII**. Perhaps here the nitroxide on the long chain **XCIX** may have been sufficiently flexible to bend back and bind to the hydrophobic cavity. These results definitely emphasize the necessity for increasing the distance of these spin label depth gages by quite small increments, otherwise anomalies such as seen with **XCVII** and **XCIX** above may be construed as significant.

Earlier, Taylor et al. and Mushak and Coleman [144, 145] had reported the interaction of the sulfonamide C with both human and bovine carbonic anhydrase B. First, a distinct difference in the binding to the human and bovine B isozymes was noted. A complete immobilization of the label was observed in the latter, whereas in the human enzyme slightly less immobilization was observed. Second, in order to examine the effects of metal ions on binding sulfonamides, they examined Co(II), Cd(II), and Hg(II), as well as the native Zn(II) forms of the human B isozyme. The similarity of the spin-labeled Co(II) enzyme and the lack of Co(II)-nitroxide spin-spin interactions confirmed both that conformation of the Co(II) and Zn(II) derivatives was quite similar, and that the nitroxide group must be 15–20 Å away from the metal ion. This identity was also shown for the bovine B isozyme substituted with Co(II) and Zn(II). With Hg(II) or Cd(II) derivatives the lack of any binding of spin label confirmed that either of these two cations at the active site precludes sulfonamide binding. In a denaturation study of the human B isozyme with urea, it was found that in 6 *M* urea, where the ORD, CD, or aromatic spectrum of the native enzyme reflected a rapid denaturation (60 min), only about 30% of the sulfonamide spin label C was released. At lower urea concentrations, progressively lower fractions of the spin label were released and were constant over long time periods. In order to unfold completely the active site, 8 *M* urea was finally required. Furthermore, the apparently stable fraction of the enzyme at lower urea concentrations could be observed by adding label *after* first exposing native enzyme to urea. This evidence supports either a stepwise unfolding mechanism or a rapid reversible equilibrium between native and unfolded forms at lower urea concentrations. Certainly, these data are another example of the monitoring of two different unfolding processes by esr and spectrophotometric methods.

In additional studies of other regions of the carbonic anhydrase molecule, Erlich et al. [140] employed two general-purpose modification reagents, the nitroxide bromoacetamide XXVIII and the maleimide XXIV (Fig. 5). After reaction of the human B isozyme with label XXVIII, a predominantly highly immobilized spectrum was observed. (A very minor narrow line component was also present, but is not discussed as it did not represent the primary reactive product of the alkylation.) This broad spectrum was not reproducible when the enzyme alkylation was carried out in the presence of acetazolamide, another strong inhibitor of carbonic anhydrase which acts similarly to sulfonamides. Since it had been previously reported that iodoacetamide inactivation was protected by an active-site inhibitor, and that this inactivation was in concert with carboxymethylation of N-3 of the imidazole of an active-site histidine, they suspected that their bromoacetamide spin label XXVIII in fact reacted analogously with this specific histidine imidazole [146]. They further suggested that the site of this histidine might be at the bottom of the active site, since its hindered rotation of the label could be rationalized strictly on steric grounds. However, a significantly hydrophobic region of the binding site could interact with the nitroxide ring sufficiently to add an attractive potential energy well in its motion.

Mushak and Coleman also looked at the bromosuccinate analog of Smith's compound [133] XXV

$$NH-\overset{\overset{\displaystyle O}{\|}}{C}-CH_2-\underset{\underset{\displaystyle Br}{|}}{C}H-CO_2H$$

XXV

with the human B isozyme [145]. They confirmed from a competive labeling experiment with iodoacetate that this derivative exclusively alkylated the specific histidine residue. They showed that the spin label modification prevented sulfonamide binding in contrast to the ineffectiveness of iodoacetate. A titration experiment with a 50%

alkylated enzyme sample and sulfonamide spin label C resulted in a composite spectrum representing 50% of the enzyme bound with bromosuccinate XXV and the other 50% with only sulfonamide label C. An examination of the x-ray model of the enzyme showed that, assuming the histidyl group was about 5 Å from the metal ion site, both labels C and XXV would occupy overlapping positions. In contrast, a small carboxymethyl group at this histidine was not a bulky steric barrier and did not inhibit sulfonamide binding.

The maleimide nitroxide XXIV is especially reactive with sulfhydryl, but less so with ε-amino (lysine), groups. The result of labeling human C isozyme with XXIV was a highly mobile spectrum indicative of unrestricted rotational freedom of the nitroxide ring (aside from those restrictions imposed by the covalent bond). Such a result usually suggests modification of lysine ε-amino groups which are almost always surface residues on a protein molecule. Although Erlich et al. [140] did not characterize this spin-labeled enzyme product further, it was suggested that the labeled site may indeed be a sulfhydryl group shown by x-ray diffraction results to be on the enzyme surface near the active site cleft. Since they had a series of sulfonamide derivatives (see Fig. 22) with which they could predict the approximate position of the nitroxide group in the enzyme, it appeared to be reasonable that the nitroxide group on maleimide label XXIV (assumed to be located at a specific surface sulfhydryl) was close enough to the paramagnetic group on label XCVI or XCVII for spin-spin interactions (see eq. 1). Unfortunately, the spectrum of a "doubly labeled" C isozyme was simply the superposition of the two component spectra. Either the two radicals were not in close juxtaposition for a sufficient period of time, or the assumed location of the maleimide label XXIV was incorrect [140].

6   APPLICATIONS TO OTHER AREAS OF BIOLOGY

Perhaps the other most important application of the spin-labeling technique in biology has been in the area of lipids and membrane structure. While this relatively new area has advanced so quickly that it deserves a chapter to review properly, it is worthwhile to describe here briefly some of the salient features. Several comprehensive reviews have been written on this subject [11, 53, 97, 147–151, 192–194].

Most of the work to date has emphasized the features of structural order and internal anisotropic motion in a membrane. If a membrane structure resembles the well-known lipid bilayer model, its constituent lipid molecules should be arranged in a fairly ordered structure with all the fatty acid chains in a parallel extended conformation. Therefore, a spin-labeled fatty acid analog such as **LXIII** incorporated in such a bilayer structure should take on an orientation dictated by its (chemically similar) environment. If all the fatty acid chains are extended, parallel, and closely packed, one should expect a spectrum of an "ordered" nitroxide molecule undergoing anisotropic rotation strictly about its long axis. The rigid oxazolidine nitroxide group on the fatty acid chain in **LXIII** insures that the nitroxide motional characteristics will be directly related to the motion of the main-chain methylene groups.

LXIII

The results to date on both artificial model membranes and real membranes have indicated that there are varying degrees of order and flexibility as one moves down into the lipid portion of the membrane from its head group [152]. These are a function, of course, of the membrane type and lipid composition. Calculations have been made on the degree of "tilt" of the observed spin-labeled lipid with respect to an ideal totally oriented system [153, 154]. For instance, evidence has been presented concerning the effects on membrane structure of bent (unsaturated) fatty acid chains [155]. The phenomenon of a flexibility gradient "down" the lipid chain in a membrane has been suggested [156]. The effects of ions, proteins, anesthetics, antibiotics, and so on, on lipid bilayer structure have been studied [148], as well as a method for determining the nature and extent of "boundary lipid" on a lipoprotein complex [157]. Calculations of the rate of flipping in and out of the membrane by the lipid label (and presumably any general fatty acid chain in the structure) have been made by measuring the rate of reduction of nitroxide by ascorbic acid (accessible only at the solvent interface), or by nmr techniques [124]. The significant problem of lateral diffusion or lateral (transport) communication in the membrane has been examined by monitoring the rate of disappearance of spin-spin interactions from some local high concentration of spin label as it begins to diffuse uniformly over the membrane [158]. For instance, the results of such a calculation indicate that a newly introduced lipid molecule would require only 1 sec to diffuse translationally to the "other side" of an *Escherichia coli* cell membrane [158].

7 OTHER EXPERIMENTAL ANGLES OF APPROACH USING SPIN LABELS

One of the most promising combinations of techniques is that of nmr and spin labels. The selective broadening of nuclear resonance lines by a paramagnet is, to an approximation, a function of the nucleus-electron separation and their relative motion. An excellent example of this technique with spin labels was described in Section 5.6. The theoretical background in this area is quite sound and has been applied to a great many paramagnetic metal ion-containing systems [119]. The spin label offers the advantage of more versatility in specific site placement. The technique of electron-electron double resonance (ELDOR) also promises to be useful in systems of (coupled) electron spins within distances such that the spin-spin term in the spin Hamiltonian becomes significant [159]. The double-resonance approach should enable a researcher to derive interlabel distances in such a system.

The pronounced effects of high microwave power (saturation) on different motional states of a nitroxide group offer us a method for selectively observing one motional state in the presence of another. The saturating (equalizing or slightly inverting the two energy-level populations so that no net microwave absorption occurs over emission) of a more mobile spin label over a more strongly immobilized one results in a spectrum in which the faster motional state has saturated and thus disappeared from the spectrum [160]. Hyde and co-workers have quite recently explored this saturation phenomenon with rotational correlation time τ_c, and have developed techniques for estimating τ_c [161]. Finally, the combination of spin labels with optical methods should also prove to be quite useful in the future. These usually yellow to orange-red molecules are characterized optically by two principal absorptions. In the UV a strong $\pi-\pi$ transition occurs at about 240 nm ($\epsilon \sim 2000$), while the visible absorption at 400–500 nm is a very weak $n-\pi*$ transition ($\epsilon \sim 5-15$) which unlike the former transition is sensitive to the solvent [25, 162]. Singer and Davis [163] showed that nitroxides can reduce the excited-state lifetime of a fluorescent chromophore by acting as a fluorescence quencher. The interaction is believed to involve excited-singlet energy transfer from the chromophore to the nitroxide [164]. The requirement is that the nitroxide have an absorption at longer wavelength than the emission band of the donor chromophore. The potential applications here are enormous, but have not been exploited significantly to date.

8 PROGNOSIS

The spin label technique is a relatively new field in physical biochemistry, having existed less than a decade. It has to its advantage several outstanding features over a comparable evolving technique of past years, such as ORD. Some obvious points are: a sound theoretical background; a relatively simple, interpretable spectrum with which one can accurately correlate physical properties and their magnitude—motion or orientation; the reporter group advantage of being the only physical observable in the system; an extreme sensitivity to a motionally hindering environment and its changes; and an extremely high absolute sensitivity (10^{-6} to 10^{-8} M).

The reporter group method also carries inherent disadvantages if not employed cautiously. Namely, the problem of structural perturbations introduced into the biomolecule by the label itself. It is especially important to emphasize this problem with respect to the spin label method, since the labor of carrying out experiments can be rather easy today with the advent of commercially available spin labels and student model esr instruments. In short, an incubation period of experience by a newly arrived researcher in this area is just as necessary as an "apprenticeship" in any other field of science.

The employment of general modifying reagents and substrate/ inhibitor analogs has provided promising new evidence about the structural features and stability of many enzymes of current biological interest. The recent combination of nmr techniques with spin labels promises to be one of the few sophisticated methods currently available for obtaining a quantitative three-dimensional picture of an enzyme in its natural environment, the physiological (aqueous) medium.

This new spectroscopic gauge of protein conformation and its relation to enzyme action is now slowly growing out of the adolescent stage. The future holds the answers to its increasing value to the biological sciences. There is no doubt that potential new pitfalls are sure to crop up in the paths of our research in this area. A continued but cautious progression in this field promises to yield significant contributions to our understanding of the fundamental structure-function problem in biochemistry.

ACKNOWLEDGMENT

The author gratefully acknowledges the research support of the National Science Foundation (GB 16437), Research Corporation, and the National Institutes of Health (GM 19364).

APPENDIX

In the excellent review by Jost and Griffith [51] an appendix was included with references to the syntheses and applications of the spin labels discussed. Since the compilation turned out to be extremely useful to the reader, a similar referencing is provided below with references to March 1973 included (see Table 6).

TABLE 6   REFERENCES TO SYNTHESES AND APPLICATIONS OF SPIN LABELS DESCRIBED IN THIS CHAPTER

Label	Reference	Label	Reference
I	30	XXII	169
II	30	XXIII	170, 171
III	32	XXIV	171, 172
IV	8, 184	XXV	133
V	162	XXVI	66
VI	165	XXVII	173
VII	8, 162	XXVIII	173, 38
VIII	166	XXIX	10, 75, 80
IX	11, 167	XXX	174
X	168	XXXI	175
XI	7, 30	XXXII	176
XII	7	XXXIII	58, 60–62
XIII	7	XXXIV	177
XIV	7	XXXV	176
XV	168	XXXVI	86, 88
XVI	7, 30	XXXVII	85, 87, 91–93
XVII	7	XXXVIII	86
XVIII	11	XXXIX	178
XIX	15	XL	141
XX	11	XLI	141
XXI	6	XLII	141

Table 6 (*Continued*)

Label	Reference	Label	Reference
XLIII	15	LXXIII	186
XLIV	15	LXXIV	189
XLV	179	LXXV	190
XLVI	123	LXXVI	187
XLVII	120, 180	LXXVII	169
XLIX	182	LXXIX	66
L	183	LXXX	66
LI	184	LXXXI	73
LII	184	LXXXII	74
LIII	184	LXXXIII	86, 88
LIV	185	LXXXIV	80
LV	185	LXXXV	86, 90
LVI	185	LXXXVI	96, 40
LVII	32	LXXXVII	87
LIX	30	LXXXVIII	86, 88
LX	51	LXXXIX	90
LXI	32	XC	116, 38
LXII	66	XCI	38
LXIII	186	XCII	38
LXIV	153	XCIII	139, 140
LXV	187	XCIV	139, 140
LXVI	51	XCV	139
LXVII	153	XCVI	139, 140
LXIX	187	XCVII	139, 140
LXX	66	XCVIII	139, 140
LXXI	153	XCIX	139, 140
LXXII	51	C	144, 145

REFERENCES

1. M. Burr and D. E. Koshland, Jr., *Proc. Nat. Acad. Sci. U.S.*, 52, 1017 (1964).
2. C. F. Meares and D. G. Westmoreland, *Cold Spring Harbor Symp. Quant. Biol.*, 36, 511 (1971).
3. D. E. Koshland, Jr., *Science*, 142, 1533 (1963).
4. T. Stone, T. Buckman, P. Nordio, and H. M. McConnell, *Proc. Nat. Acad. Sci. U.S.*, 54, 1010 (1965).
5. S. Ohnishi and H. M. McConnell, *J. Am. Chem. Soc.*, 87, 2293 (1965).
6. A. K. Hoffman and A. T. Henderson, *J. Am. Chem. Soc.*, 83, 4671 (1961).
7. E. G. Rozantsev and L. A. Krinitskaya, *Tetrahedron*, 21, 491 (1965).

8. E. G. Rozantsev and M. B. Nieman, *Tetrahedron*, **20**, 131 (1964).
9. The many publications of A. Rassat; a principal reference is *Bull. Soc. Chim. Fr.*, **1965**, 3273 (1965).
10. C. L. Hamilton and H. M. McConnell, in *Structural Chemistry and Molecular Biology*, A. Rich and N. Davidson, Eds., W. H. Freeman, San Francisco, 1968, p. 115.
11. H. M. McConnell and B. G. McFarland, *Quart. Rev. Biophys.*, **3**, 91 (1970).
12. L. J. Libertini and O. H. Griffith, *J. Chem. Phys.*, **53**, 1359 (1970).
13. W. L. Hubbell and H. M. McConnell, *Proc. Nat. Acad. Sci. U.S.*, **64**, 20 (1969).
14. J. Seelig, *J. Am. Chem. Soc.*, **92**, 3881 (1970).
15. J. C. Hsia and L. H. Piette, *Arch. Biochem. Biophys.*, **129**, 296 (1969).
16. G. E. Pake *Paramagnetic Resonance*, W. A. Benjamin, New York, 1962, Chap. 5.
17. J. H. Freed, G. V. Bruno, and C. Polnaszek, *J. Phys. Chem.*, **75**, 3385 (1971).
18. I. V. Alexandrov, A. N. Ivanova, N. N. Korst, A. V. Lazarev, A. I. Prikhozhenko, and V. R. Stryukov, *Mol. Phys.*, **18**, 681 (1970).
19. M. S. Itzkowitz, *J. Chem. Phys.*, **46**, 3048 (1967).
20. E. J. Shimshick and H. M. McConnell, *Biochem. Biophys. Res. Commun.*, **46**, 321 (1972).
21. R. C. McCalley, E. J. Shimshick, and H. M. McConnell, *Chem. Phys. Letters*, **13**, 115 (1972).
22. R. P. Haugland and L. Stryer, *Conformation Biopolymers*, **1**, 321 (1967).
23. O. H. Griffith and P. C. Jost, personal communication.
24. A. D. Keith and R. J. Melhorn, personal communication.
25. T. Kawamura, S. Matsunami, and T. Yonezawa, *Bull. Chem. Soc. Japan*, **40**, 1111 (1967).
26. M. Bersohn and J. C. Baird, *An Introduction to Electron Paramagnetic Resonance*, W. A. Benjamin, New York, 1966.
27. A. Carrington and A. D. McLachlan, *Introduction to Magnetic Resonance*, Harper and Row, New York, 1964.
28. J. Wertz and J. R. Bolton, *Electron Spin Resonance—Elementary Theory and Applications*, McGraw-Hill, New York, 1972.
29. See also G. R. Luckhurst and G. F. Pedulli, *J. Am. Chem. Soc.*, **92**, 4738 (1970).
30. E. G. Rozantsev, *Free Nitroxyl Radicals*, Plenum Press, New York, 1970.
31. See also A. R. Forrester, J. M. Hay, and R. H. Thomson, *Organic Chemistry of Stable Free Radicals*, Academic Press, New York, 1968, Chap. 5.
32. J. F. W. Keana, S. B. Keana, and D. Beetham, *J. Am. Chem. Soc.*, **89**, 3055 (1967).
33. F. Francis, *J. Chem. Soc.*, 1927, 2897.
34. E. G. Rozantsev, *Izv Akad. Nauk. SSSR Ser. Khim.*, **1966**, 1770.
35. Chapter 4 of reference 30.

36. A. Horwitz, Ph.D. Thesis, Stanford University, Stanford, Calif., 1970.

37. J. D. Morrisett and H. R. Drott, *J. Biol. Chem.*, 244, 5083 (1969).

38. R. W. Wien, J. D. Morrisett, and H. M. McConnell, *Biochemistry*, 11, 3707 (1972).

39. L. J. Berliner and T. Frey, unpublished results.

40. S. S. Wong, K. Quiggle, C. Triplett, and L. J. Berliner, *J. Biol. Chem.*, 249, 1678 (1974).

41. H. Stratigus, unpublished results.

42. R. Aneja and A. P. Davies, *Chem. Phys. Lipids*, 4, 60 (1970).

43. J. C. A. Boeyens and G. J. Kruger, *Acta Crystallogr.*, B26, 668 (1970).

44. W. Turley and F. P. Boer, *Acta. Crystallogr.*, B28, 1641 (1972).

45. L. Pauling, *The Nature of the Chemical Bond*, Cornell University Press, Ithaca, N.Y., 1960, p. 344.

46. P. J. Lajzérowicz-Bonnetau, *Acta Crystallogr.*, B24, 196 (1968).

47. L. J. Berliner, *Acta Crystallogr.*, B26, 1198 (1970).

48. P. A. Capiomont, B. Chion, and J. Lajzérowicz, *Acta Crystallogr.*, B27, 322 (1971).

49. P. A. Capiomont, *Acta Crystallogr.*, B28, 2298 (1972).

50. S. S. Ament, J. B. Wetherington, J. W. Moncrief, K. Flohr, M. Mochizuki and E. T. Kaiser, *J. Am. Chem. Soc.*, 95, 7896 (1973).

51. P. Jost and O. H. Griffith, *Methods in Pharmacology*, Vol. II, C. Chignell, Ed., Appleton-Century-Crofts, New York, 1972, p. 223.

52. H. M. McConnell and J. C. A. Boeyens, *J. Phys. Chem.*, 71, 12 (1967).

53. I. C. P. Smith, D. Marsh, and S. Schreier-Mucillo, in *Free Radicals in Molecular Pathology*, Vol. I, W. A. Pryor, Ed., Academic Press, New York 1974.

54. I. C. P. Smith, in *Biological Applications of Electron Spin Resonance Spectroscopy*, H. Swartz, J. R. Bolton, and D. Borg, Eds., Wiley, New York, 1972, p. 483.

55. O. H. Griffith and A. S. Waggoner, *Accounts Chem. Res.*, 2, 17 (1969).

56. S. Ohnishi, *Seibutsu Butsuri*, 8, 118 (1968).

57. J. D. Ingham, *J. Macromol. Sci.-Rev. Macromol. Chem.*, C2, 279 (1968).

58. L. J. Berliner and H. M. McConnell, *Proc. Natl. Acad. Sci. U.S.*, 55, 708 (1966).

59. See M. L. Bender and F. J. Kézdy, *Ann. Rev. Biochem.*, 34, 49 (1965).

60. K. Flohr and E. T. Kaiser, *J. Am. Chem. Soc.*, 94, 3675 (1972).

61. K. Flohr, R. M. Paton, and E. T. Kaiser, *Chem. Commun.*, 1621 (1971).

62. L. J. Berliner and H. M. McConnell, *Biochem. Biophys. Res. Commun.*, 43, 651 (1971).

63. D. M. Blow, M. G. Rossmann, and B. A. Jeffery, *J. Mol. Biol.*, 8, 65 (1964).

64. T. A. Steitz, R. Henderson, and D. M. Blow, *J. Mol. Biol.*, 46 337 (1969).

65. S. S. Wong and L. J. Berliner, unpublished results.

66. D. J. Kosman, J. C. Hsia, and L. H. Piette, *Arch. Biochem. Biophys.*, 133, 29 (1969).

67. S. G. Cohen, L. H. Klee, and S. Y. Weinstein, *J. Am. Chem. Soc.*, **88**, 5302 (1966).
68. S. G. Cohen and J. Crossley, *J. Am. Chem. Soc.*, **86**, 4999 (1964).
69. R. Henderson. *J. Mol. Biol.*, **54**, 341 (1970).
70. D. E. Koshland, Jr., D. H. Strumeyer, and W. J. Ray, Jr., *Brookhaven Symp. Biol.*, **15**, 101 (1962).
71. J. R. Knowles, *Biochem. J.*, **95**, 180 (1965).
72. J. T. Gerig, *J. Am. Chem. Soc.*, **90**, 2681 (1968).
73. L. J. Berliner, Ph.D. Thesis, Stanford University, Stanford, Calif., 1967.
74. D. J. Kosman and L. H. Piette, in *Magnetic Resonances in Biological Systems*, C. Franconi, Ed., Gordon and Breach, New York, 1971, p. 287.
75. D. J. Kosman and L. H. Piette, *Arch. Biochem. Biophys.*, **149**, 452 (1972).
76. B. H. Landis and L. J. Berliner, to be published.
77. B. W. Matthews, P. B. Sigler, R. Henderson, and D. M. Blow, *Nature*, **214**, 652 (1967).
78. H. L. Oppenheimer, B. Labouesse, and G. P. Hess, *J. Biol. Chem.*, **241**, 2720 (1966).
79. G. P. Hess, J. McConn, E. Ku, and G. McConkey, *Phil. Trans. Roy. Soc: London*, **B257**, 89 (1970).
80. D. J. Kosman, *J. Mol. Biol.*, **67**, 247 (1972).
81. D. M. Blow, in *The Enzymes*, Vol. III, P. Boyer, Ed., Academic Press, New York, 1971, Chap. 6.
82. For a discussion and review see G. P. Hess, in *The Enzymes*, Vol. III, P. Boyer, Ed., Academic Press, New York, 1971, Chap. 7.
83. E. B. Ong, E. Shaw, and G. Schoellmann, *J. Biol. Chem.*, **240**, 694 (1965).
84. H. Weiner, W. N. White, D. G. Hoare, and D. E. Koshland, Jr., *J. Am. Chem. Soc.*, **88**, 3851 (1966).
85. J. D. Morrisett, C. A. Broomfield, and B. E. Hackley, Jr., *J. Biol. Chem.*, **244**, 5756 (1969).
86. J. C. Hsia, D. J. Kosman, and L. H. Piette, *Biochem. Biophys. Res. Commun.*, **36**, 75 (1969).
87. J. D. Morrisett and C. A. Broomfield, *J. Am. Chem. Soc.*, **93**, 7297 (1971).
88. J. C. Hsia, D. J. Kosman, and L. H. Piette, *Arch. Biochem. Biophys.*, **149**, 441 (1972).
89. For a discussion of biradical nitroxide spectral interpretation, see G. R. Luckhurst and G. F. Pedulli, *J. Am. Chem. Soc.*, **92**, 4738 (1970).
90. L. J. Berliner, *Biochemistry*, **11**, 2921 (1972).
91. L. J. Berliner and S. S. Wong, *J. Biol. Chem.*, **248**, 1118 (1973).
92. J. D. Morrisett and C. A. Broomfield, *J. Biol. Chem.*, **247**, 7224 (1972).
93. A. J. Hoff, R. A. Oosterbaan, and R. Deen, *FEBS Letters*, **14**, 17 (1971).
94. Staffan Magnusson, in *The Enzymes*, Vol. III, P. Boyer, Ed., Academic Press, New York, 1971, Chap. 9.
95. D. Nachmansohn and I. B. Wilson, in *Methods in Enzymology*, Vol. I., S. P. Colowick and N. O. Kaplan, Eds., Academic Press, New York, 1955, p. 642.

96. L. J. Berliner and S. S. Wong, *J. Biol. Chem.*, 249, 1668 (1974).

97. L. J. Berliner (Ed.), *Spin Labeling—Theory and Applications*, Academic Press, New York, 1974.

98. D. E. Fahrney and A. M. Gold, *J. Am. Chem. Soc.*, 85, 997 (1963).

99. A. M. Gold and D. E. Fahrney, *Biochemistry*, 3, 783 (1964).

100. R. M. Stroud, L. M. Kay and R. E. Dickerson, *Cold Spring Harbor Symp. Quant. Biol.*, 36, 125 (1971).

101. A. R. Fersht and Y. R. Requena, *J. Am. Chem. Soc.*, 93, 7079 (1971).

102. C. J. Martin and G. M. Bhatnagar, *Biochemistry*, 5, 1230 (1966).

103. F. Friedberg, J. E. Long, and A. S. Brecher, *Proc. Soc. Exp. Biol. Med.*, 130, 1046 (1969).

104. T. R. Hopkins and J. D. Spikes, *Biochem. Biophys. Res. Commun.*, 28, 480 (1967).

105. R. Goldman, L. Goldstein, and E. Katchalski, in *Biochemical Aspects of Reactions on Solid Supports*, G. Stark, Ed., Academic Press, New York, 1971, Chap. 1.

106. G. P. Royer and John P. Andrews, *J. Biol. Chem.*, 248, 1807 (1973).

107. L. J. Berliner, S. M. Miller, R. Uy, and G. P. Royer, *Biochem. Biophys. Acta*, 315, 195 (1973).

108. G. P. Royer and R. Uy, *J. Biol. Chem.*, 248, 2627 (1973).

109. B. D. Gabel, I. Z. Steinberg, and E. Katchalski, *Biochemistry*, 10, 4661 (1971).

110. R. M. Paton and E. T. Kaiser, *J. Am. Chem. Soc.*, 92, 4723 (1970).

111. M. Hennrich and F. Cramer, *J. Am. Chem. Soc.*, 87, 1121 (1965).

112. M. L. Bender, R. L. Van Etten, G. A. Clowes, and J. F. Sebastian, *J. Am. Chem. Soc.*, 88, 2318, 2319 (1966); 89, 3242, 3253 (1967).

113. C. C. F. Blake, G. A. Mair, A. C. T. North, D. C. Phillips, and V. R. Sarma, *Proc. Roy. Soc. London*, B167, 365 (1967).

114. C. C. F. Blake, L. N. Johnson, G. A. Mair, A. C. T. North, D. C. Phillips, and V. R. Sarma, *Proc. Roy. Soc. London*, B167, 378 (1967).

115. F. N. Dahlquist, T. Rand-Meir, and M. A. Raftery, *Proc. Nat. Acad. Sci. U.S.*, 61, 1194 (1968).

116. L. J. Berliner, *J. Mol. Biol.*, 61, 189 (1971).

117. J. A. Rupley, *Proc. Roy. Soc. London*, B167, 248 (1967).

118. J. A. Rupley, L. Butler, M. Gerring, F. J. Hartdegen, and R. Pecoraro, *Proc. Nat. Acad. Sci. U.S.*, 57, 1088 (1967).

119. A. S. Mildvan and M. Cohn, *Advan. Enzymol.*, 33, 1 (1970).

120. A. S. Mildvan and H. Weiner, *J. Biol. Chem.*, 244, 2465 (1969).

121. T. R. Krugh, *Biochemistry*, 10, 2594 (1971).

122. A. Bennick, I. D. Campbell, R. A. Dwek, N. C. Price, G. K. Radda, and A. G. Salmon, *Nature New Biol.*, 234, 140 (1971).

123. G. C. K. Roberts, J. Hannah, and O. Jardetzky, *Science*, 165, 504 (1969).

124. R. D. Kornberg and H. M. McConnell, *Biochemistry*, 10, 1111 (1971).

125. H. Sternlicht and E. Wheeler, in *Magnetic Resonance in Biological Systems*, A. Ehrenberg, B. G. Malmstrom, and T. Vanngard, Eds., Pergamon Press,

Oxford, 1967, p. 325. Also see comment on this paper by H. M. McConnell, I. C. P. Smith, and N. Bhacca, p. 335.

126. H. W. Wyckoff, K. D. Hardman, N. M. Allewell, T. Inagami, L. N. Johnson, and F. M. Richards, *J. Biol. Chem.*, 242, 3749, 3984 (1967).

127. H. A. Scheraga and J. A. Rupley, *Advan. Enzymol.*, 24, 161 (1962).

128. W. H. Stein, *Federation Proc.*, 23, 599 (1964).

129. A. M. Crestfield, W. H. Stein, and S. Moore, *J. Biol. Chem.*, 238, 2413, 2421 (1963).

130. R. L. Heinrikson, *J. Biol. Chem.*, 241, 1393 (1966).

131. R. L. Heinrikson, W. H. Stein, A. M. Crestfield, and S. Moore, *J. Biol. Chem.*, 240, 2921 (1965).

132. C. H. W. Hirs, M. Halmann, and J. H. Kycia, in *Biological Structure and Function*, Vol. I., T. W. Goodwin and I. Lindberg, Eds., Academic Press, New York, 1961, p. 41.

133. I. C. P. Smith, *Biochemistry*, 7, 745 (1968).

134. W. F. Harrington and J. A. Schellman, *Compt. Rend. Trav. Lab. Carlsberg, Ser. Chim.*, 30, 21 (1956).

135. L. Li, J. P. Riehmand, and H. A. Scheraga, *Biochemistry*, 5, 2043 (1966).

136. S. Kalman, K. Linderström-Lang, M. Ottesen, and F. M. Richards, *Biochim. Biophys. Acta*, 16, 297 (1955).

137. S. J. Paterson, H. Dugas, and I. C. P. Smith, unpublished results.

138. J. D. Morrisett, Ph.D. Thesis, University of North Carolina, 1969.

139. C. F. Chignell, D. K. Starkweather, and R. H. Erlich, *Biochem. Biophys. Acta*, 271, 6 (1972).

140. R. H. Erlich, D. K. Starkweather, and C. F. Chignell, *Mol. Pharmacol.*, 9, 61 (1973).

141. J. C. Hsia and L. H. Piette, *Arch. Biochem. Biophys.*, 132, 466 (1969).

142. P. C. Bergstén, I. Waara, S. Lövgren, A. Liljas, K. K. Kannan, and U. Bengston in *Oxygen Affinity of Hemoglobin and Red Cell Acid Base Status*, Alfred Benzon Symposium IV, Munksgaard, Copenhagen, 1972, p. 363.

143. A. Liljas, K. K. Kannan, P. C. Bergsten, I. Waara, K. Fridborg, B. Strandberg, U. Carlbom, L. Järup, S. Lövgren, and M. Petef, *Nature New Biol.*, 235, 131 (1972).

144. J. S. Taylor, P. Mushak, and J. E. Coleman, *Proc. Nat. Acad. Sci. U.S.*, 67, 1410 (1970).

145. P. Mushak and J. E. Coleman, *J. Biol. Chem.*, 247, 373 (1972).

146. P. L. Whitney, P. O. Nyman, and B. G. Malmstrom, *J. Biol. Chem.*, 242, 4212 (1967).

147. P. Jost, A. S. Waggoner and O. H. Griffith, in *Structure and Function of Biological Membranes*, L. Rothfield, Ed., Academic Press, New York, 1971, p. 83.

148. I. C. P. Smith, *Chimia*, 25, 349 (1971).

149. S. Schreier-Muccillo and I. C. P. Smith, in *Progress in Surface and Membrane Science*, Vol. 9, D. A. Cadenhead, J. F. Danielli, and M. D. Rosenberg, Eds., Academic Press, New York, 1974.

150. B. G. McFarland, in *Methods in Enzymology—Biomembranes*, S. Fleischer, L. Packer, and R. Estabrook, Eds., Academic Press, New York, Vol. 33, 1974.

151. R. J. Melhorn and A. D. Keith, in *Molecular Biology of Membranes*, C. F. Fox and A. D. Keith, Eds., Sinauer Associates, Stamford, Conn., 1972, p. 192.

152. W. L. Hubbell and H. M. McConnell, *J. Am. Chem. Soc.*, 93, 314 (1971).

153. P. C. Jost, L. J. Libertini, V. Hebert, and O. H. Griffith, *J. Mol. Biol.*, 59, 77 (1971).

154. R. D. Lapper, S. J. Paterson and I. C. P. Smith, *Can. J. Biochem.*, 50, 969 (1972).

155. B. G. McFarland and H. M. McConnell, *Proc. Nat. Acad. Sci. U.S.*, 68, 1274 (1971).

156. H. M. McConnell and B. G. McFarland, *Trans. N.Y. Acad. Sci.*, 195, 207 (1972).

157. P. C. Jost, O. H. Griffith, R. A. Capaldi, and G. Vanderkooi, *Proc. Nat. Acad. Sci. U.S.*, 70, 480 (1973).

158. For a short review see H. M. McConnell, P. Devaux, and Carl Scandella, *Membrane Res.*, C. F. Fox (Ed.), Academic Press, New York, 1972, p. 27.

159. J. S. Hyde, J. C. W. Chien, and J. Freed, *J. Chem. Phys.*, 48, 4211 (1968).

160. J. Gergely, in *Fifth International Conference on Magnetic Resonance in Biological Systems*, New York, Dec. 1972, *Annals N.Y. Acad. Sci.*, 222, 574 (1973).

161. J. S. Hyde and D. D. Thomas, *ibid.* 222, 680 (1973).

162. R. Briére, H. Lemaire, and A. Rassat, *Bull. Soc. Chim. Fr.*, 1965, 3273.

163. L. A. Singer and G. A. Davis, *J. Am. Chem. Soc.*, 89, 158 (1967).

164. A. L. Buchachenko, M. S. Khloplyankina, and S. N. Dobryakov, *Opt. Spektrosk. (USSR)*, 22, 554 (1967).

165. E. G. Rozantsev, V. A. Golubev, and M. B. Nieman, *Izv. Akad. Nauk, SSSR, Ser. Khim.* 1965, 391.

166. E. G. Rozantsev and Y. V. Kokhanov, *Izv. Akad. Nauk. SSSR, Ser. Khim.*, 8, 1477 (1966).

167. Y. Tonomura, S. Watanabe, and M. Morales, *Biochemistry*, 8, 2171 (1969).

168. A. Rassat and P. Rey, *Bull Soc. Chim. Fr.*, 1967, 815.

169. A. D. Keith, A. S. Waggoner, and O. H. Griffith, *Proc. Nat. Acad. Sci. U.S.*, 61, 819 (1968).

170. O. H. Griffith and H. M. McConnell, *Proc. Nat. Acad. Sci. U.S.*, 55, 8 (1966).

171. M. D. Barratt, A. P. Davies, and M. T. A. Evans, *Eur. J. Biochem.*, 24, 280 (1971).

172. S. Ohnishi, J. C. A. Boeyens, and H. M. McConnell, *Proc. Nat. Acad. Sci. U.S.*, 56, 809 (1966).

173. S. Ogawa and H. M. McConnell, *Proc. Nat. Acad. Sci. U.S.*, 58, 19 (1967).

174. B. M. Hoffman, P. Schofield, and A. Rich, *Proc. Nat. Acad. Sci. U.S.*, 62, 1195 (1969).

175. J. C. A. Boeyens and H. M. McConnell, *Proc. Nat. Acad. Sci. U.S.*, **56**, 22 (1966).

176. O. H. Griffith, J. F. W. Keana, D. L. Noall, and J. L. Ivey, *Biochem. Biophys. Acta*, **148**, 583 (1967).

177. O. H. Griffith, J. F. W. Keana, S. Rottschaefer, and T. A. Warlick, *J. Am. Chem. Soc.*, **89**, 5072 (1967).

178. L. Stryer and O. H. Griffith, *Proc. Nat. Acad. Sci. U.S.*, **54**, 1785 (1965).

179. A. S. Waggoner, O. H. Griffith, and C. R. Christensen, *Proc. Nat. Acad. Sci. U.S.*, **57**, 1198 (1967).

180. H. Weiner, *Biochemistry*, **8**, 526 (1969).

181. R. T. Ogata and H. M. McConnell, *Cold Spring Harbor Symp. Quant. Biol.*, **36**, 325 (1972).

182. D. Kabat, B. Hoffmann, and A. Rich, *Biopolymer* **9**, 95 (1970).

183. Z. Ciecierska-Tworek, S. P. Van, and O. H. Griffith, *J. Mol. Struct.*, **16**, 139 (1973).

184. R. Briére, R. -M. Dupeyre, H. Lemaire, C. Morat, A. Rassat, and P. Rey, *Bull. Soc. Chim. Fr.*, **1965**, 3290.

185. P. Ferruti, D. Gill, M. P. Klein, H. H. Wang, G. Entene, and M. Calvin, *J. Am. Chem. Soc.*, **92**, 3704 (1970).

186. W. L. Hubbell and H. M. McConnell, *Proc. Nat. Acad. Sci. U.S.*, **64**, 20 (1969).

187. W. L. Hubbell and H. M. McConnell, *Proc. Nat. Acad. Sci. U.S.*, **63**, 16 (1969).

188. A. S. Waggoner, T. J. Kingzett, S. Rottschaefer, O. H. Griffith, and A. D. Keith, *Chem. Phys. Lipids*, **3**, 245 (1969).

189. J. C. Hsia, H. Schneider, and I. C. P. Smith, *Chem. Phys. Lipids*, **4**, 238 (1969).

190. A. S. Waggoner, A. D. Keith, and O. H. Griffith, *J. Phys. Chem.*, **72**, 4129 (1968).

191. J. C. Lee, R. P. Frigon, J. Hirsh, J. Thomas, and S. N. Timasheff, *Federation Proc.*, **32**, 496 (1973), Abstr. 1547.

192. A. D. Keith, M. Sharnoff, and G. E. Cohn. *Biochim. Biophys. Acta*, *(Biomembranes Rev.)* 344(MR3), 1974.

193. A. D. Keith and M. A. Williams, in *Methods in Enzymology— Biomembranes*, S. Fleischer, L. Packer, and R. Estabrook, Eds., Academic Press, New York, Vol. 33, 1974.

194. A. D. Keith, B. J. Wisnieski, S. Henry, and J. C. Williams, in *Lipids and Biomembranes of Eukaryotic Microorganisms*, J. A. Erwin, Ed., Academic Press, New York, 1973, p. 259.

195. J. B. Wetherington, S. S. Ament and J. W. Moncrief, *Acta. Cryst.*, **B30**, 568 (1974).

196. W. B. Gleason, *Acta. Cryst.*, **B 29**, 2959 (1973)..

197. M. Krieger, L. M. Kay and R. M. Stroud, *J. Mol. Biol.*, **82**, 209 (1974).

"HALF-OF-THE SITES" REACTIVITY AND THE ROLE OF SUBUNIT INTERACTIONS IN ENZYME CATALYSIS

M. LAZDUNSKI

Centre de Biochimie, Université de Nice, Nice, France

1 INTRODUCTION

Our knowledge of enzyme structures has improved considerably in the last decade. Spectacular advances have been made in the fields of sequence elucidation, determination of subunit structures, x-ray crystallography, and electron microscopy of proteins. They have taught us that a very large number, probably a vast majority, of intracellular enzymes are multisubunit structures. Subunits are often identical, and association in most cases does not appear to involve covalent bonds. A very characteristic property of multisubunit enzymes is their high degree of symmetry (Table 1). For all these reasons, investigation of the functional importance of the subunit structure of enzymes is of primary interest.

Present views [22, 23] concerning the reasons why molecular evolution has so frequently favored the appearance and maintenance of polymeric and symmetric enzymes can be briefly summarized.

1. One of the most general advantages, according to Monod et al. [22], is "to decrease the surface-volume ratio as well as to cover up the hydrophobic areas of the monomers. It is evidently more economical to achieve such results, whenever possible, by associating monomers rather than by increasing the unit molecular weight (i.e., the molecular weight per active center)."

2. More subtle advantages follow from cooperative effects that are a natural consequence of a symmetric structure and which have been used to increase the efficiency of respiratory carriers and of feedback control systems.

3. Such structures introduce the possibility of combining chemically different subunits with different preformed specificities so that either their mutual catalytic efficiency (e.g., in two stages of a complex reaction) is enhanced, or one ligand can regulate

TABLE 1 SYMMETRY OF SOME OLIGOMERIC ENZYMES[a]

Enzymes	Number of Subunits	Symmetry Class[b]	Reference
Escherichia coli alkaline phosphatase	2	D_1 (X)	1
Liver alcohol dehydrogenase	2	D_1 (X)	2
Muscle malate dehydrogenase	2	D_1 (X)	3
Muscle triose phosphate isomerase	2	D_1 (X)	4
Muscle phosphoglucose isomerase	2	D_1 (X)	5
Yeast hexokinase	2	D_1 (X)	6
Muscle lactate dehydrogenase	4	D_2 (X)	7
Muscle glyceraldehyde-3-phosphate dehydrogenase	4	D_2 (X)	8
Yeast phosphoglycerate mutase	4	D_2 (X)	9
Muscle aldolase	4	D_2 (X) (EM)	10, 11
Muscle phosphorylase b	4	D_2 (X, EM)	
Tryptophanase	4	D_2 (EM)	
Glutamate dehydrogenase	6	D_3 (EM)	12, 13
Escherichia coli aspartate transcarbamylase	6 catalytic 6 regulatory	D_3 (X)	14
Glutamine synthetase	12	D_6 (X, EM)	17, 18

[a] This table is not exhaustive; references for further examples will be found in other reviews [19–21]. Multienzyme complexes such as fatty acid synthetase or the pyruvate dehydrogenase complex have also been shown to be highly symmetric arrangements.
[b] X, X-ray crystallography; EM, electron microscopy.

allosterically the metabolism of another ligand of different structure. Well-known examples in these two categories are tryptophan synthetase [24] and aspartate transcarbamylase [25].

Allosteric enzymes that are characterized by a sigmoidal response to increasing substrate or inhibitor concentration have been extensively studied in recent years. Monod et al. [22] and Koshland et al. [26] showed that cooperative kinetics for these enzymes necessitate polymeric structures with indirect interactions between distinct binding sites. These interactions are mediated by structural changes [22, 26]. The stereochemical details of these structural changes are now well described in the case of hemoglobin [27]. For more details concerning allosteric enzymes, the reader is referred to several recent excellent reviews [28–31].

Actually, most of the enzymes known to be oligomeric are not known to be allosteric. This is fairly reasonable, since most polymeric proteins are probably not endowed with specific regulatory functions. In fact most enzymes appear to display Michaelis-Menten kinetics rather than sigmoidal v versus S profiles.

While there is an obvious and important functional meaning of the subunit structure of allosteric enzymes, the most immediate interpretation of classic Michaelian kinetics with polymeric enzymes is that the different active sites on each constitutive monomer function independently. This interpretation seems to be generally accepted, but it fails to indicate why there might be an important functional advantage in oligomeric structures.

In fact, there is an ever-increasing amount of literature describing negative (or anti-) cooperativity of substrate binding or substrate transformation by polymeric enzymes obeying Michaelian kinetics. Negative cooperativity indicates that constituent protomers do not behave as independent entities in oligomeric structures. To try to resolve the apparent paradox between anticooperative binding, or half-of-the-sites reactivity, and the classic Michaelian kinetics, a new interpretation has been proposed recently concerning the functional interrelationship between distinct active sites on identical monomers that are part of oligomeric enzymes obeying Michaelian kinetics [32].

The theoretical aspects of negative cooperative interactions between distinct binding sites are not treated here. Very detailed and interesting descriptions of this subject have been recently published by Koshland and his group [26, 30, 31, 33, 34].

In this review, our interest is centered on kinetic cooperativity that is the basis of flip-flop-type mechanisms, and on the analysis of some of the recent literature concerning enzymes in which only half of the active sites appear to react.

2. SUBUNIT INTERACTIONS IN ENZYME CATALYSIS. DESCRIPTION OF A MODEL, ALKALINE PHOSPHATASE

Studies carried out with *Escherichia coli* and intestinal alkaline phosphatases are described in some detail. This chapter serves to illustrate the essential properties of a flip-flop mechanism.

2.1 Subunit Structure and the Nature of the Active Site of Alkaline Phosphatase of *E. Coli*

Alkaline phosphatase is a dimeric enzyme with a molecular weight of 86,000. There is only one structural gene for the protein, indicating that the enzyme is composed of identical subunits [35, 36]. Alkaline phosphatase is a zinc metalloenzyme [37]. The apoenzyme is devoid of catalytic activity, but Zn^{2+} can be replaced by a variety of metal ions such as Co^{2+}, Cd^{2+}, Mn^{2+}, Cu^{2+}, or Ni^{2+}. Studies concerning Zn^{2+} [38, 39], Co^{2+} [39, 40], Cd^{2+} [41, 42], Mn^{2+} [42, 43], and Cu^{2+} [44] phosphatases have now shown that there are four essential metal binding sites on the enzyme. They belong to two different families; there appear to be two tight and two loose metal ion-binding sites. This has been demonstrated in a variety of ways; the method using the electon paramagnetic resonance (epr) technique is shown in Figure 1.

The metal atom not only plays a fundamental role in the catalytic mechanism of alkaline phosphatase, but also plays a structural role [40, 45, 46]. Figure 2 shows that the dimeric Zn^{2+}- phosphatase is considerably more resistant to alkaline dissociation into monomers than the apophosphatase.

Schwartz [47] and Engström [48] showed that the enzyme's catalytic mechanism involves the intermediate formation of a phosphoryl enzyme. The critical site of phosphorylation in the active center is a serine residue. The phosphatase catalytic mechanism can then be schematized:

$$E + ROPO_3^{2-} \rightleftharpoons ES \longrightarrow \underset{\underset{PO_3^{2-}}{\overset{|}{O}}}{\overset{|}{E}} \longrightarrow E + PO_4H^{2-}$$

$$+ ROH$$

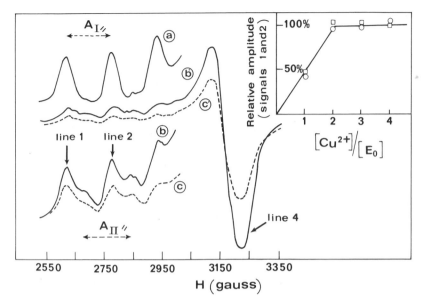

Figure 1   Epr spectra. (a) Apophosphatase plus two gram atoms of copper per mole of protein. (b and b′) Apophosphatase plus four gram atoms of copper per mole of protein. (c and c′) Spectrum of the 1 : 1 complex formed between this Cu^{2+} phosphatase ($[Cu^{2+}]/[E_0]$ = 4) and inorganic phosphate. Inset: Cu^{2+}/apoprotein ratio dependence for the intensities of low-field lines 1 and 2. From reference 44.

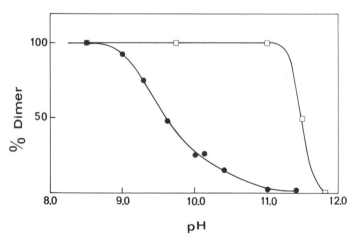

Figure 2   pH dependence of alkaline phosphatase stability. The percentage of dimer at alkaline pH is measured by ultracentrifugation after a 15-hr incubation at 30°C. $[E_0]$ = 6 mg/ml; Zn^{2+} alkaline phosphatase with 4 zinc atoms per mole (□); apoenzyme in 1 μM EDTA (●).

The serine residue and the zinc atoms are presently the only essential elements that have been clearly demonstrated to be at the active site. Chemical modifications of the enzyme with acetic anhydride, acetylimidazole, and tetranitromethane [49, 50] have failed to inactivate the enzyme. In consequence, in the present state of our knowledge, lysine residues and phenol side chains do not seem to be involved in the catalytic process.

pK_m–pH curves for the Zn^{2+}- and also for the Co^{2+} enzyme (Fig. 3) indicate that the acidic form of an ionizable group with a pK_a of 8.6 at 25°C in the Zn^{2+} enzyme is important for the catalytic process [51]. This pK_a is 8.9 in the Co^{2+} enzyme [52]. The observed pK_a may represent that of a metal-coordinated water molecule. A coordinated water molecule has also been implicated in the catalytic mechanism of another metalloenzyme, carbonic anhydrase [53, 55].

2.2 A Characteristic Property of the Enzyme: Negative Cooperativity and Half-Site Reactivity

Phosphorylation of the strategic serine residue can be performed either with the substrate, as indicated in the preceding paragraph, or with the reaction product PO_4H^{2-}.

$$E + PO_4H^{2-} \rightleftharpoons E \cdot PO_4H^{2-} \rightleftharpoons E–OPO_3^{2-}$$

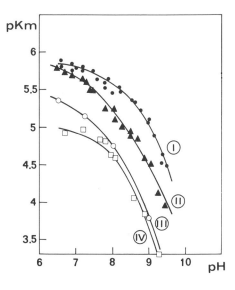

Figure 3 pK_m–pH profiles obtained with Zn^{2+} alkaline phosphatase. The substrates are: p-nitrophenyl phosphate (I) at 25°C and (II) at 45°C; β-glycerophosphate (III) at 45°C; and glucose 1-phosphate (IV) at 45°C. From reference 51.

Figure 4 shows the extent of phosphorylation of the enzyme with high concentrations of orthophosphate at different pH values [56]. The extent of phosphorylation is higher at acidic pH; this observation was first made many years ago by Engström [48] and by Schwartz [47], and was also recently confirmed by Wilson and his group [57, 58] and by Applebury and Coleman [59]. Zn^{2+} alkaline phosphatase incorporates covalently 2 moles of phosphate per mole of enzyme at acidic pH, and none at alkaline pH.

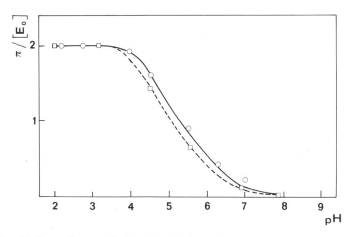

Figure 4 pH dependence of phosphorylation of the active centers of alkaline phosphatase of *E. coli* with [^{32}P] orthophosphate (10 m*M*). Zn^{2+} phosphatase at $0°C$ (□); Co^{2+} phosphatase at $23°C$ (○). π/E_0 = moles of [^{32}P] phosphate covalently bound (π) per mole of enzyme (E_0). From reference 32.

The noncovalent binding of inorganic phosphate to metallo-phosphatases at alkaline pH was studied using a variety of techniques [32, 42, 43, 50, 56, 60] including equilibrium dialysis, gel filtration, spectrophotometry, and epr spectroscopy. The characteristic feature of the binding is anticooperativity.

Our equilibrium studies have shown the adsorption of 2 moles of phosphate per mole of enzyme. However, in spite of the identity of the peptide chains in the subunits, the sites are not independent. One site binds phosphate tightly, and saturation occurs at a low phosphate concentration. The saturation of the second site is much more difficult and requires much higher concentrations of phosphate. A typical representation of this anticooperative behavior is presented in Figure 5. Noncovalent complexes containing 2 moles

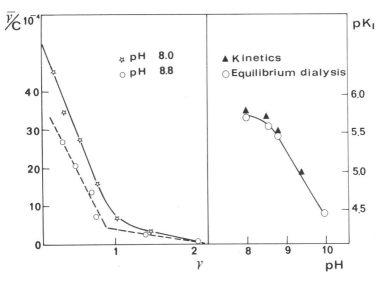

Figure 5 (Left) Scatchard plots for the binding of orthophosphate to Zn^{2+} phosphatase. $\bar{\nu}$, mole ratio of binding P_i per dimer; c, equilibrium concentration of orthophosphate; 25°C, 0.4 M NaCl. (Right) Variation of pK_I with pH (K_I: dissociation constant of the enzyme-orthophosphate complex). ▲, Kinetic determinations: ○, pK_I was calculated from the slope $(1/K_I)$ of the $\bar{\nu}/c$ versus $\bar{\nu}$ plots for the saturation of the first (tight) site. From reference 56.

of orthophosphate per mole of Zn^{2+} phosphatase have been isolated [56]. They are stable only at high protein concentrations; they dissociate at lower concentrations into 1 : 1 complexes which in turn dissociate on further dilution to regenerate the free enzyme. Complexes containing only 1 mole of orthophosphate per mole of Co^{2+} or Cu^{2+} phosphatase have also been prepared [56, 44]. A typical representation showing the enzyme concentration dependence of the stability of noncovalent complexes formed with orthophosphate and Zn^{2+} or Cu^{2+} phosphatase appears in Figure 6. A concentration of 4-5 μM of Zn^{2+} phosphatase is sufficient to prepare the complex with only one bound phosphate. A concentration higher than 25 μM is necessary to obtain the complex with two noncovalently bound phosphates. The anticooperativity is stronger for Cu^{2+} phosphatase, since only the 1:1 complex can be isolated even at protein concentrations as high as 30 μM. Absolute negative cooperativity was also found for the binding of arsenate to Zn^{2+} phosphatase [61].

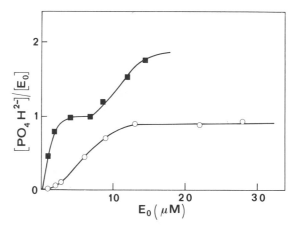

Figure 6   Enzyme concentration dependence of the stability of the noncovalent [^{32}P] orthophosphate-phosphatase complex. Zn^{2+} phosphatase (■); Cu^{2+} phosphatase (○). The enzyme-orthophosphate complex was formed by incubation of high concentrations of Zn^{2+} or Cu^{2+} phosphatase (7–30 mg/ml) with radioactive orthophosphate (10 mM). The complex was isolated on Sephadex G-25 equilibrated at pH 8.0. After isolation the complex was analyzed for radioactivity. The same operation was repeated at different dilutions of the isolated complex. From reference 44.

This anticooperativity also exists in the formation of covalent derivatives at acidic pH. Figure 7 presents the concentration dependence for the covalent phosphorylation of the active centers of Zn^{2+} and Co^{2+} phosphatases with orthophosphate at pH 4.2 and 5.0. The experimental data are in accordance with the calculated curves for the phosphorylation of two different sites with two distinct Michaelis constants K_1 and K_2. At pH 4.2 the values of these constants are 25 μM and 1 mM for Zn^{2+} phosphatase, and 25 μM and 0.5 mM for Co^{2+} phosphatase. K_1 and K_2 differ by a factor of 40 for the Zn^{2+} enzyme, and a factor of 20 for Co^{2+} phosphatase. This ratio corresponds to an apparent free energy of interaction of 2.1 kcal mole^{-1}. The anticooperative character of the covalent phosphorylation is even more dramatic at pH 5.0. At an ionic strength of 0.05, the values of the Michaelis constants are $K_1 = 10$ μM and $K_2 = 7.0$ mM; K_1 and K_2 differ by a factor of 700. In this case the free energy of interaction is 3.8 kcal mole^{-1}. These data indicate that the anticooperative character of the covalent binding does exist, and that it increases with increasing pH.

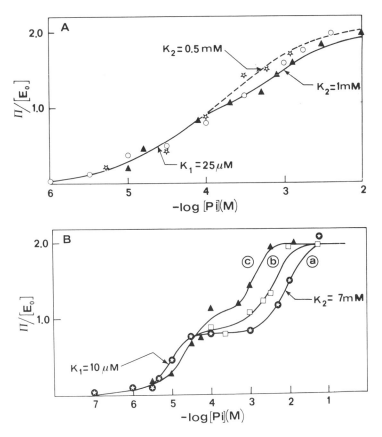

Figure 7 The $[^{32}P]$ orthophosphate concentration dependence of the covalent phosphorylation of Zn^{2+} and Co^{2+} phosphatase at $20°C$ [32]. (A) pH 4.2, ionic strength 0.05; ———, Zn^{2+} phosphatase, ---, Co^{2+} phosphatase. The solid and broken lines which pass through the experimental points are theoretical curves assuming two active sites with different affinities for orthophosphate (B) Behavior of Zn^{2+} phosphatase at pH 5.0, $20°C$, ionic strength 0.05 (*a*), 0.15 (*b*), 0.5 (*c*).

In Figure 8 kinetic evidence is shown for anticooperativity. The diphosphorylated derivative

$$E\begin{array}{c}\nearrow P\\\searrow P\end{array}$$

is first formed by incubation of the enzyme with high concentrations of unlabeled phosphate. A pulse of [^{32}P] orthophosphate is added to the mixture at time 0, and the exchange of unlabeled phosphate for

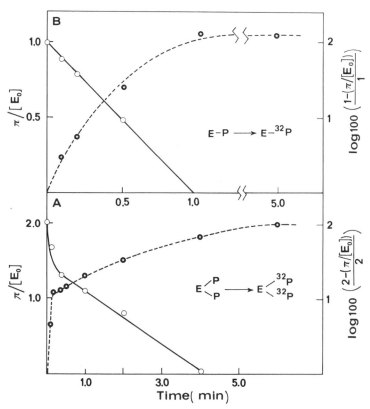

Figure 8   The kinetics of the turnover of phosphorylated derivatives of Zn^{2+} phosphatase at pH 4.2, 20°C. The exchange of the covalent phosphate of diphosphorylated $(E<\begin{smallmatrix}P\\P\end{smallmatrix})$ and of monophosphorylated (E−P) phosphatases (A and B). The π/E_0 versus time plots are presented as broken lines. The logarithmic plots are shown by solid lines.

[^{32}P] orthophosphate is measured. One phosphate is exchanged rapidly, and the second one is exchanged more slowly. Complete exchange apparently proceeds through two successive first-order processes. The same experiment has been done for the monophosphorylated derivative E—P. In that case, the exchange is a pure first-order reaction. Similar data have been obtained with the Cu^{2+} enzyme [44].

Orthophosphate is the reaction product in the hydrolysis of substrates by alkaline phosphatase. Up to now, we have dealt with equilibrium conditions. Figure 9 presents a picture of the steady

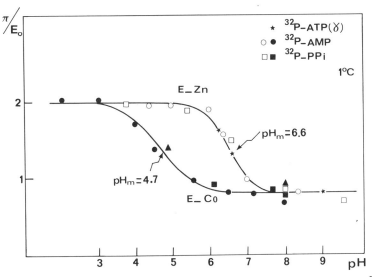

Figure 9   pH dependence of phosphorylation of the active centers of Zn^{2+} and Co^{2+} alkaline phosphatases with substrates in the steady state. Maximal phosphorylation of Zn^{2+} phosphatase with [^{32}P] AMP (○), [^{32}P] ATP (*), and [^{32}P] pyrophosphate (□). Maximal phosphorylation of Co^{2+} phosphatase with [^{32}P] AMP (●), [^{32}P] pyrophosphate (■), and p-nitrophenyl phosphate (▲). From reference 32.

state at different pH values after quenching the phosphorylated derivative formed from organophosphates and pyrophosphates. Substrates were chosen for their differences in chemical reactivity. They are [^{32}P] ATP, [^{32}P] AMP, [^{32}P] pyrophosphate, and p-nitrophenyl phosphate. Two moles of phosphate can be incorporated covalently at acidic pH, showing that in this pH range inorganic phosphate and substrates behave the same way. The situation is quite different at alkaline pH, since instead of no labeling with orthophosphate one observes the covalent incorporation of about 1 mole of phosphate per mole of enzyme with all substrates. This observation strongly suggests that under steady-state conditions only one site is phosphorylated at any instant at alkaline pH where the enzyme normally operates. The second important observation is that despite their different chemical reactivity all substrates give the same steady-state kinetics.

Two moles of phosphate per mole of Zn^{2+} phosphatase can be incorporated at acidic pH using high concentrations of substrates (Figure 9). Again, the two sites do not behave as if they were equivalent and independent. The anticooperative character of the covalent phosphorylation at acidic pH has been observed with AMP (Figure 10) as was observed with inorganic phosphate (Figure 2). At pH 5.0 only one site is phosphorylated at low concentrations of AMP (10–100 μM), and phosphorylation of the other site requires much higher concentrations of the substrate (about 10 mM).

Another approach to the demonstration of half-of-the sites reactivity for alkaline phosphatase is the analysis of transient kinetics. Fernley and Walker [62], Ko and Kézdy [63], and Trentham and Gutfreund [64] observed the liberation of only 1 mole of 4-methylumbelliferone [62], or of 2,4-dinitrophenol [63, 64], per mole of enzyme in the presteady-state phase of the reaction

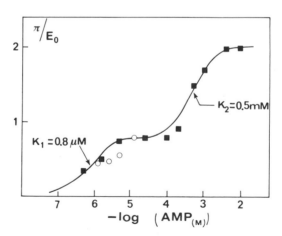

Figure 10 The [^{32}P] AMP concentration dependence of the covalent phosphorylation of Zn^{2+} phosphatase under steady-state conditions at pH 5.0. From reference 32.

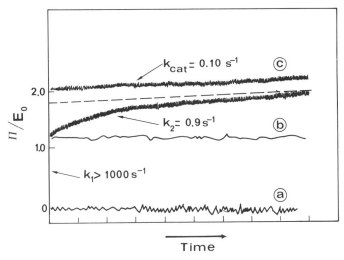

Figure 11 Stopped-flow oscilloscope tracing measuring absorbance changes related to 2.4 dinitrophenol appearance during Co^{2+} phosphatase hydrolysis of 2.4 dinitrophenyl phosphate. $[E_0] = 3.5\ \mu M$, $[S_0] = 10^{-3}\ M$, pH 4.20, $5°C$. a, Base line, 5 msec per division; b presteady-state phase, 5 msec per division; c, presteady-state and steady-state phases, 0.5 sec per division. k_1 and k_2 are first-order rate constants describing transient kinetics. $k_{cat} = V_m/[E_0]$.

between Zn^{2+} phosphatase and 4-methylumbelliferone phosphate or 2,4-dinitrophenyl phosphate at pH 5.5–6.0. All these investigators used low substrate concentrations (less than 10 μM). This result is in agreement with the demonstration that, in the same pH range and with low substrate concentrations, only about 1 mole of phosphate is incorporated covalently per mole of phosphatase (Figure 10).

A stopped-flow analysis of the transient kinetics of the phosphatase-catalyzed hydrolysis of 2,4-dinitrophenyl phosphate has also been carried out at 10°C and "high" substrate concentration by Chappelet in this laboratory in collaboration with Iwatsubo. The burst of 2,4-dinitrophenol liberated in the presteady-state phase varies considerably with pH. A burst of about 2 moles of dinitrophenol per mole of enzyme is observed at pH 4.20 (Figure 11). It is a biphasic burst, which again indicates the nonequivalence of the sites. The first site reacts very rapidly ($k_1 > 1000\ sec^{-1}$) and gives an instant burst of 1 mole of 2-4 dinitrophenol per mole of dimer; the second site reacts much more slowly ($k_2 = 0.9\ sec^{-1}$).

TABLE 2 COMPARATIVE KINETIC PROPERTIES OF

Substrate	Zn^{2+} Phosphatase[a]	
	V_m (μmoles $\times$ min^{-1} $\times$ mg^{-1})	$10^5\ K_m\ (M)$
p-Nitrophenyl phosphate	28.0	0.27
β-Glycerophosphate	28.6	0.61
AMP	26.7	0.61
ATP	21.5	0.77
PP$_i$	28.0	0.61
Phosphoenol pyruvate	28.0	1.47

[a] pH 8.4, 25°C, 0.4 M NaCl.

[b] pH 8.4, 25°C, 0.32 M Na$_2$SO$_4$.

2.3 Characteristic Catalytic Properties of the Enzyme: Michaelis-Menten Kinetics and Lack of Specificity

Although negative cooperativity clearly indicates that the two sites of the alkaline phosphatase of E. coli are not independent, the enzyme obeys the Michaelis-Menten law at alkaline pH over a very wide range of substrate concentration (0.4 μM to 10 mM) [32]. This is the typical kinetic behavior of monomeric enzymes. In consequence, Michaelis-Menten kinetics suggests independence of the sites in the catalytic process.

A unique property of the alkaline phosphatase is its lack of specificity. With Zn^{2+} alkaline phosphatase, the V_m and K_m values are very similar if not identical for most substrates, in spite of considerable differences in chemical reactivity (Table 2). The maximal activity of Co^{2+} phosphatase is lower than that of Zn^{2+} phosphatase by a factor of about 10 at alkaline pH [49, 52], whereas the maximal activity of Cu^{2+} phosphatase is lower by a factor of about 100 [44]. The K_m values also vary somewhat among Zn^{2+}, Co^{2+}, and Cu^{2+} phosphatase. There are other differences among these

Zn^{2+}, Co^{2+}, AND Cu^{2+} PHOSPHATASE [73, 76]

Co^{2+} Phosphatase[a]		Cu^{2+} Phosphatase[b]	
V_m (μmoles $\times$ min^{-1} $\times$ mg^{-1})	10^5 K_m (M)	V_m (μmoles $\times$ min^{-1} $\times$ mg^{-1})	10^5 K_m (M)
2.9	0.7	0.31	0.8
2.7	1.7	0.28	1.8
2.9	1.9	0.43	2.4
2.6	2.4	0.60	2.4
		0.50	2.8
2.8	4.0	0.27	3.5

metalloenzymes, such as V_m versus pH profiles, ionic strength, and nucleophile effects [44, 49, 52]. In spite of all these differences, all three metallophosphatases share the unusual property that their V_m and K_m values for all substrates are very similar or identical (Table 2). The observation that the rate of enzymic hydrolysis of a family of phosphomonoesters is independent of the chemical nature of the leaving group, which may seem hard to explain at first sight in terms of known mechanisms of phosphate ester hydrolysis, is of a vital importance for bacteria. Alkaline phosphatase is synthetized by *E. coli* cells only when inorganic phosphate becomes limiting in the culture medium [65]. The function of the enzyme is then obviously related to the ability of the bacteria to grow on organic phosphate or pyrophosphate as the sole source of inorganic phosphate. Then, as soon as the enzyme is formed under limiting phosphate conditions, it must hydrolyze its substrates as rapidly and as completely as possible for the bacterial cell to resume normal growth. The situation would obviously be complicated if the different substrates had widely different V_m and K_m values. A substrate with particularly low K_m

and V_m values could act as a competitive inhibitor for the hydrolysis of other substrates. An efficient process to provide the bacteria with inorganic phosphate implies that all substrates have similar or identical V_m and K_m values.

2.4 Description of a Flip-Flop-Type Mechanism for the Action of Alkaline Phosphatase of *E. Coli*

The mechanism proposed for the catalytic activity of alkaline phosphatase of *E. coli* relates two apparently antinomic properties of the enzyme, namely, Michaelian kinetics and negative cooperativity. It also explains the lack of specificity of the catalysis.

Figure 12 presents the proposed mechanism [32]. The alkaline phosphatase is presented as symmetric in the free state, F; in such a model the subunits are functionally identical. Five different geometries of the active site are indicated in the minimum model: two different geometries of the free state, F and F_i (F binds the substrate easily, while F_i binds it only with difficulty); the T form of the active site is stabilized by reversible (noncovalent) association with the substrate, while the R and R_i species exist only in the phosphorylated state (R_i does not dephosphorylate; R does dephosphorylate).

In step 1, the binding of one phosphate molecule (orthophosphate or substrates) induces a conformational change (or displaces a preequilibrium). The enzyme molecule now appears asymmetric with two different geometries of the active site. At this point the enzyme can follow two different paths. It can either bind a second substrate molecule noncovalently (step 6) to form a new symmetric T–T complex (horizontal pathway), or undergo the phosphorylation step (step 2) which involves a structural rearrangement, T becoming R_i (right subunit), while the free active site F_i (left subunit) regains the ability to bind a new substrate molecule easily (step 3). Because of anticooperativity, complexes containing two noncovalent phosphates per mole of enzyme are not favored at alkaline pH, and the horizontal pathway (step 6) is not followed.

Step 3 results in the formation of a T–R dimer which contains 1 mole of phosphate covalently bound to the R species, and 1 mole of substrate noncovalently bound to the T species. This is the really important complex which accumulates at alkaline pH under steady-state conditions when the enzyme acts upon its substrates.

Step 5 (horizontal pathway) does not take place at alkaline pH, because of the very anticooperative character of the covalent phosphorylation. The enzyme follows step 4, the heart of the catalytic process, which we call the flip-flop step. In this step, there is a transformation of the T—R species into an R—T species which is its mirror image. One site phosphorylates, while the other dephosphorylates and binds a new substrate molecule. Step 4 is the slowest step at alkaline pH for substrates such as AMP, ATP, p-nitrophenyl phosphate, pyrophosphate, or any other substrate hydrolyzed at the same maximal rate. Step 4 is in fact made up of three elementary steps which are represented in Figure 12B. The slowest elementary step is the dephosphorylation. This point is discussed in the legend for the figure. It is not known whether the rate of dephosphorylation is controlled by the chemical reaction or by the conformational change.

The rate-limiting step for inorganic phosphate is step 2. k_4, the rate constant of step 4 is 40 sec^{-1} for organic esters at pH 9 [51], whereas k_2, the rate constant of step 2 for orthophosphate is 0.4–1.6 sec^{-1} at the same pH. Because step 2 is the rate-limiting step for inorganic phosphate, the reaction product at high concentrations can form 2:1 noncovalent complexes with the enzyme, thus stabilizing the T—T species at alkaline pH. The pathway beginning with step 5 is of no importance at alkaline pH, but is very important at acidic pH. In this pH range, step 5 (k_5) becomes faster than step 4, the flip-flop step. Covalent anticooperativity decreases, and dephosphorylation of the diphosphorylated derivative (k_{-5}) becomes rate-limiting at acidic pH with high concentrations of substrate or inorganic phosphate.

It is now necessary to summarize the salient evidence in favor of such a mechanism.

All but one of the complexes involved in the flip-flop mechanism have been identified and isolated.

They contain one or two covalent or one or two noncovalent phosphates per mole of protein. The only complex that obviously could not be isolated is the central one which contains both one covalent and one noncovalent phosphate per mole of enzyme and which accumulates under steady-state conditions with inorganic phosphates or pyrophosphates as substrates. For identification, this complex had to be quenched at a very acidic pH at which the noncovalently bound molecule of substrate is eliminated by denaturation. However, it has been well established that this complex contains 1 mole of covalent phosphate per mole of protein (Figure 5). This demonstrates the reactivity of half of the sites. Furthermore, the following data suggest that a hybrid complex containing both one covalently bound and one noncovalently bound arsenate

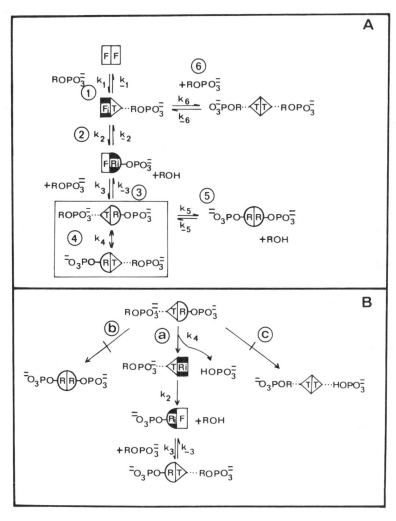

Figure 12 (A) The flip-flop mechanism of the alkaline phosphatase of *E. coli* [32]. F, F_i, R, R_i, and T represent five different geometries of the active site. $\boxed{R}\text{-}OPO_3^{2-}$ or $\boxed{R_i}\text{-}OPO_3$ represents covalent binding of phosphate to the serine residue of the active site. $T\!\!>\cdots ROPO_3^{2-}$ represents noncovalent binding. $ROPO_3^{2-}$ can be either the substrate (AMP, ATP, pyrophosphate, etc.) or the product of the hydrolysis, orthophosphate ($HOPO_3^{2-}$). $ROPO_3^{2-}$ represents the ionization state of the substrates and of orthophosphate at alkaline pH. At acidic pH, this state should be replaced in the figure by the monoanionic or the neutral forms.

(B) The anatomy of the flip-flop step. Once the hybrid complex (arsenate, a competitive inhibitor of the enzyme, is an analog of phosphate) has been isolated at pH 8.5 using Co^{2+} alkaline phosphatase [61].

A Co^{2+} phosphatase-(arsenate)$_2$ complex is easily prepared using enzyme concentrations higher than 20 μM; it dissociates into the stoichiometric Co^{2+} phosphatase-(arsenate)$_1$ complex when the enzyme concentration drops to about 5 μM. Table 3 shows that gel-filtration chromatography of the Co^{2+} phosphatase-([^{74}As]-arsenate)$_2$ complex through a Sephadex column equilibrated with unlabeled arsenate results in the exchange of only one of the two radioactive arsenate molecules bound to the protein. The other arsenate molecule is so strongly bound to the Co^{2+} enzyme that it does not exchange during chromatography. This strongly suggests covalent association of one of the two arsenates with the enzyme. When the noncovalent (loosely bound and easily exchangeable) arsenate is first removed from the 2 : 1 complex by simple dilution, long enough incubation of the resulting and presumably covalent Co^{2+} phosphatase-([^{74}As] arsenate)$_1$ permits dearsenylation and easy replacement of radioactive arsenate by unlabeled arsenate (Table 3).

The flip-flop mechanism involves the negative cooperativity found both for noncovalent binding of the substrate and for covalent phosphorylation. This negative cooperativity is a necessary property of flip-flop-type catalysis.

The simple steady-state kinetic treatment of the flip-flop mechanism at alkaline pH indicates that the alternate phosphorylation and dephosphorylation of the sites leads to pure Michaelis-Menten behavior [32]. The phenomenological approach to flip-flop-type mechanisms is described later.

$ROPO_3^{2-} \cdots \langle T|R \rangle -OPO_3^{2-}$ is formed, two possibilities arise: the phosphorylation of the T geometry or the dephosphorylation of the R geometry. Phosphorylation of the T geometry, (pathway b) would give the diphosphorylated derivative which in fact does not form at alkaline pH (covalent anticooperativity). This alternative should then be excluded. Dephosphorylation of the R state can occur either via pathway a or c. Pathway c will lead to the T–T species with 1 mole of orthophosphate noncovalently bound per mole of enzyme. Such a species is stable only at high orthophosphate concentrations at alkaline pH (noncovalent anticooperativity), and should not be taken into consideration for initial rate kinetics. Pathway a is the only possibility. Dephosphorylation of the R geometry results in the automatic exclusion of $HOPO_3^{2-}$, the product of the reaction. k_4 is the rate constant of the slowest step with substrates, AMP, ATP, etc.

TABLE 3 STABILITY OF THE RADIOACTIVE COMPLEXES FORMED BETWEEN [^{74}As] ARSENATE, [^{32}P] ORTHOPHOSPHATE, AND ALKALINE PHOSPHATASE[a]

Type of Complex before Exchange	Conditions of Exchange	Residual Bound [^{32}P] Phosphate and [^{74}As] Arsenate (moles per mole)
Zn^{2+} Phosphatase–(^{32}P] ortho-phosphate)$_2$	Chromatography at pH 8 on Sephadex G-25 equilibrated with 10 mM unlabeled orthophosphate	0.0
Co^{2+} Phosphatase–([^{74}As] arsenate)$_2$	Chromatography at pH 8 on Sephadex G-25 equilibrated with 10 mM unlabeled arsenate	0.85
Co^{2+} Phosphatase–([^{74}As] arsenate)$_1$[a]	Incubation 1 min in 0.1 mM unlabeled arsenate and chromatography at pH 8 on Sephadex G-25 equilibrated with 10 mM Tris-Cl	0.0

[a] This complex was prepared by dilution of the Co^{2+} phosphatase–([^{74}As] arsenate)$_2$ complex; after isolation, and before the exchange experiment with unlabeled arsenate, it was kept for 3 weeks at 4°C and pH 7.7.

The functional identity of the two subunits implies that symmetry exists in the alkaline phosphatase of *E. coli*. This has been recently demonstrated by x-ray studies on single crystals; the monomer is the asymmetric unit [1].

The flip-flop mechanism implies that the structure of the alkaline phosphatase remains dimeric during the course of the catalysis. Molecular-weight estimations of the enzyme in the presence of organic and inorganic phosphates have shown that this hypothesis is well founded [52]. Also, no active monomer of the enzyme obtained through mutation [66, 67], culture in the presence of amino acid analogs [68], or acidic [69] or alkaline dissociation has ever been demonstrated.

The flip-flop mechanism involves, in the first step, a conformational change in the active site, which is triggered by the noncovalent binding of 1 mole of substrate per mole of enzyme. The dimeric structure then loses its symmetry. The conformational change in the subunit that remains free is indicated by anticooperativity. Moreover, several indirect lines of evidence indicate an overall structural rearrangement. When orthophosphate is bound, one observes perturbations in the visible spectra of Co^{2+} [39, 70] and of Cu^{2+} phosphatase [44], of the optical rotatory dispersion (ORD) and circular dichroism (CD) spectra in Co^{2+} phosphatase [39, 70] and in the epr spectra of Cu^{2+} and Mn^{2+} phosphatase [42, 44, 71]. They probably reflect a structural reorganization of the chelate structure of the active site.

Typical perturbations in the visible and CD spectra of Co^{2+} phosphatase and in the epr spectrum of Cu^{2+} and Mn^{2+} phosphatase are presented in Figures 1 and 13–15.

Stopped-flow analysis indicates that perturbation in the visible spectrum of Co^{2+} phosphatase, following binding of inorganic phosphate or β-glycerophosphate, occurred in less than 2.5 msec (Figure 15).

An inhibitor-induced conformational change in Zn^{2+} alkaline phosphatase has also been recently observed using 2-hydroxy-5-nitrophenyl phosphonate [72, 73]. The spectrum of this phosphonate is changed upon binding to the enzyme. The kinetics of the reaction was followed by stopped-flow and temperature-jump techniques. A unimolecular rearrangement of the enzyme-inhibitor complex occurs with a rate constant of 5.4 sec^{-1}. These data constitute excellent evidence for a structural change in the active site after binding of a substrate analog.

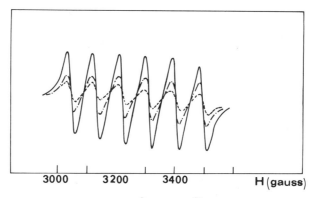

Figure 13   Epr spectra at pH 8.0, 20°C, of Mn^{2+} phosphatase (apoenzyme plus four gram atoms of manganese per mole of protein) (–), of this Mn^{2+} phosphatase plus 1 equivalent of inorganic phosphate (– • – •), and plus 2 equivalents of inorganic phosphate (----). From reference 42.

The possibility exists, however, that the R group of the phosphonate $(R-PO_3{}^{2-})$ participates in binding to alkaline phosphatase (as a result of the presence of the unesterified phenol function), whereas R groups of other phosphonates and of substrates are not recognized by the enzyme. In favor of this interpretation is the observation of an important perturbation in the phosphonate

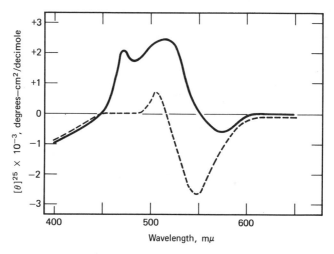

Figure 14   CD of Co^{2+} phosphatase (–) and the Co^{2+} phosphatase–orthophosphate complex (----) at pH 8.0. From reference 39.

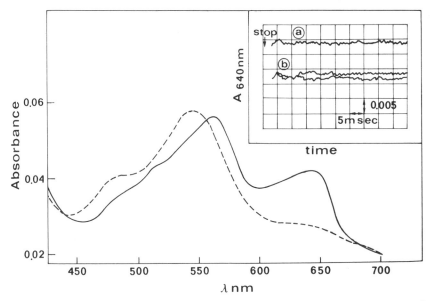

Figure 15 The influence of inorganic phosphate on the spectrum of Co^{2+} phosphatase (apoenzyme plus four gram atoms of cobalt per mole of protein). Co^{2+} phosphatase $(1.25 \times 10^{-4} \ M)$ $(-)$; Co^{2+} phosphatase (1.25×10^{-4}) plus inorganic phosphate $(2.5 \times 10^{-4} \ M)$ $(---)$. pH 7.8, 20°C. Similar results were first obtained by Vallee [39] and Coleman [70]. Inset: Stopped-flow oscilloscopic tracings of the spectral change at 640 nm; pH 7.8, 5°C. a, Co^{2+} phosphatase alone $(0.65 \times 10^{-4} \ M)$; b, Co^{2+} phosphatase $(0.65 \times 10^{-4} \ M)$ mixed with 1 or 2 equivalents of inorganic phosphate.

spectrum upon binding to the enzyme, and the fact that K_I, the dissociation constant of the complex between Zn^{2+} phosphatase and 2-hydroxy-5-nitrophenyl phosphonate is 30 μM, whereas K_I values for phosphite or phenyl phosphonate are higher than 1 mM [32]. In consequence the rate of isomerization of the enzyme-phosphonate complex might not be directly related to the rate of isomerization of classic enzyme-substrate complexes.

One of the important activity features of the flip-flop mechanism for alkaline phosphatase activity is that dephosphorylation of one of the active sites is dependent upon the noncovalent binding of a second substrate molecule to the other site (Figure 12). This is demonstrated in Figure 16. The monophosphorylated derivative is stable at high concentrations (one order of magnitude higher than the dissociation constant of the complex), but dilution permits slow

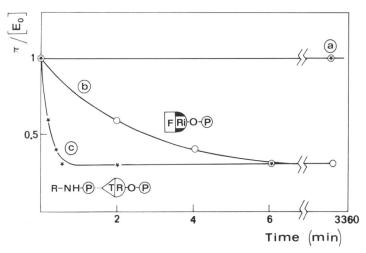

Figure 16 The stability of the monophosphorylated derivative of the Zn^{2+} alkaline phosphatase [32]; pH 5.0, 0°C, 0.4 M NaCl. a, The stability of the monophosphorylated phosphatase at a concentration of 0.1 mM (dissociation constant of the complex 0.01 mM). b, The dephosphorylation of the monophosphorylated derivative at a concentration of 0.5 μM. c, The dephosphorylation of the monophosphorylated derivative at a concentration of 0.1 mM in the presence of 1.7 mM p-chloroanilido phosphonate. From reference 32.

dephosphorylation. Dephosphorylation is accelerated by a factor of 10 when the free site is saturated with p-chloroanilido phosphonate, a substrate analog for alkaline phosphatase.

There are two consecutive chemical steps in the catalytic mechanism of alkaline phosphatase: phosphorylation and dephosphorylation of the active site. The flip-flop mechanism couples each one of these chemical steps to the binding of one molecule of substrate. There is both intrasubunit and intersubunit kinetic cooperativity. In the first two steps of the flip-flop mechanism, phosphorylation is dependent upon the conformational change that occurs on binding the first substrate molecule. Intrasubunit "activation" is necessary for the phosphorylation step. The phosphorylated subunit then needs a second activation which occurs via intersubunit interaction upon noncovalent binding of a second substrate molecule to the free site. Intrasubunit kinetic cooperativity permits phosphorylation; intersubunit kinetic cooperativity permits dephosphorylation. The situation is summarized in Figure 17 more simply than in Figure 12.

One of the steps in this mechanism, $-E-E- + S \rightarrow -E-E^*-S$ appears *only* in the transients and *not* under steady-state conditions. This observation has interesting experimental implications. For example, let us take a situation in which dephosphorylation is limiting under steady-state conditions but in which the step outside the cycle is much slower than dephosphorylation. Under such conditions, no burst of ROH is seen in the transients, and one is tempted to conclude that phosphorylation is limiting, which is not true. Conversely, quenching of the enzymic complex formed under steady-state conditions, after the enzyme has made several turns in the cycle (during 5 sec), permits one to isolate the monophosphoryl enzyme and to reach the right conclusion, that is, dephosphorylation is rate-limiting. Such a situation may well occur in the case of *E. coli* alkaline phosphatase. Analysis of transient kinetics at alkaline pH and 20–25°C indicates no burst of ROH [62, 64], whereas the enzymic complex that accumulates under steady-state conditions in the same pH range is the monophosphorylated derivative [32]. The reason for the slowness of the extracycle substrate binding step is

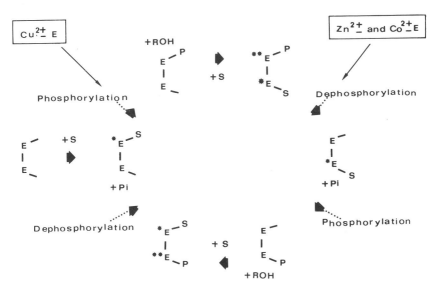

Figure 17 A simple representation of the flip-flop mechanism of alkaline phosphatase activity. *E and **E indicate intra and intersubunit activations, respectively. E—S represents noncovalent binding of the substrate, whereas *E—P or E—P represent covalent phosphorylation. Activation is a prerequisite both for phosphorylation (*E) and dephosphorylation (**E).

that association of the substrate is accompanied by displacement of a slow preequilibrium between two structural forms of the alkaline phosphatase $-E-E- \rightarrow -E-E^*-$. The existence of such a preequilibrium, presumably from a symmetric to an assymetric form, has been recently demonstrated by Halford [73].

In the mechanism shown in Figure 17, there are three types of catalytically important conformational changes:

1. Displacement of a preequilibrium which favors substrate binding to only one of the subunits.

2. A conformational change occurring after substrate binding and associated with phosphorylation of the active site; this rearrangement, which occurs during transfer from the metal to the serine residue permits association of a second substrate molecule with the free site.

3. A conformational change associated with dephosphorylation to return to the $-E-E-S$ complex.

Under our experimental conditions with Zn^{2+} and Co^{2+} phosphatase acting on organic phosphates or pyrophosphate, dephosphorylation was the rate-limiting step in the flip-flop mechanism. However, this is by no means an essential condition; phosphorylation could as well be rate-limiting. No covalent phosphorylation occurs with inorganic phosphate at alkaline pH as shown in Figure 4, but a turnover of $[^{18}O]$ orthophosphate has been demonstrated in the past [47], meaning that phosphorylation (see Figure 17) is limiting in this case.

Chemical modifications of the active site may be expected to bring changes in the rate-limiting steps. The zinc atoms play a fundamental role in the mechanism of the enzyme's activity. The apoenzyme has no catalytic activity and is unable to form noncovalent complexes or phosphorylated derivatives when incubated with substrates or inorganic phosphate [56]. The replacement of zinc by any other metal, such as cobalt, copper, manganese, or cadmium, does not eliminate catalytic activity completely.

Mn^{2+}, Cu^{2+}, and Cd^{2+} phosphatases have very low activity, but they preserve the capacity for phosphorylation and dephosphorylation [32, 42, 46]. Noncovalent as well as covalent anticooperativity persists after the replacement of zinc atoms. This property has been observed not only for Zn^{2+} phosphatase, but also for the Co^{2+}, Cu^{2+}, and Cd^{2+} enzymes [32, 44, 56]. Dephosphorylation is the rate-limiting step in the flip-flop mechanism of both Zn^{2+} and the

Co^{2+} phosphatase with substrates at alkaline pH, while phosphorylation (see Fig. 17) is rate-limiting for the Cu^{2+} enzyme. Cu^{2+}
phosphatase can not be phosphorylated at alkaline pH with AMP,
pyrophosphate, or orthophosphate [44], and gives no burst of
2,4-dinitrophenol in stopped-flow kinetics with 2,4-dinitrophenyl
phosphate. The replacement of zinc by cobalt or copper does not
change the rate-limiting step (phosphorylation) with orthophosphate
at alkaline pH [32].

From the mechanism shown in Figure 17, it is expected that any
modification in the enzyme structure or in the substrate structure
that could favor symmetric noncovalent or covalent complexes,
S–E–E–S or P–E–E–P, will take the phosphatase out of the
normal steady-state cycle. As a result, the rate of enzymic catalysis
should be considerably decreased. Figure 18 shows that Mn^{2+}
phosphatase forms extremely stable noncovalent complexes containing 2 moles of phosphate per mole of phosphatase. No phosphate
dissociation was observed at enzyme concentrations as low as 2 μM.
The activity of Mn^{2+} phosphatase is negligible as compared to that of
Zn^{2+} or Co^{2+} phosphatase. Cd^{2+} phosphatase forms covalent
complexes at alkaline pH. Figure 19 shows that at low
concentrations of AMP (1 mM) only one site is phosphorylated at pH
8.0. Both sites can be phosphorylated with high concentrations of

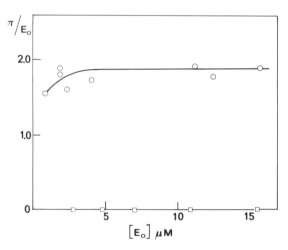

Figure 18 Enzyme concentration dependence of the total amount of
[^{32}P] inorganic phosphate bound per mole of Mn^{2+} phosphatase (○). No binding
was observed with apophosphatase (□). pH 7.6, 25°C. From reference 43.

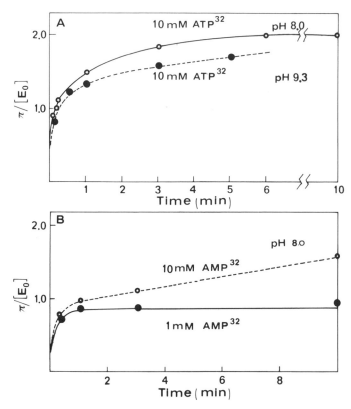

Figure 19 Time dependency of the covalent phosphorylation of Cd^{2+} phosphatase with $[^{32}P]$ ATP and $[^{32}P]$ AMP. From reference 32.

substrates such as ATP or AMP. Under such conditions, negative cooperativity persists; one site is phosphorylated very rapidly, but the second site is phosphorylated at a much slower rate. This means that replacement of Zn^{2+} by Cd^{2+} favors the formation of P–E–E–P from –E–E–P, but not that of S–E–E–P. Consequently, it is not surprising that the activity of Cd^{2+} phosphatase is considerably lower (at least about 1000 times) than that of Zn^{2+} phosphatase. The situation encountered with Cd^{2+} phosphatase at alkaline pH is similar to that observed with Zn^{2+} phosphatase at acidic pH (Figure 7), where catalytic activity is also negligible as compared to that observed at alkaline pH.

Several changes in the substrate structure may also bring a decrease in catalytic efficiency. For example, *E. coli* alkaline phosphatase hydrolyzes the oxygen esters of phosphorothioates at a rate 100- to 200-fold slower than the analogous oxyphosphates at alkaline pH [74]. Stopped-flow analysis of the transients shows that in this case, as in the two previous ones, a monophosphorylated derivative of the phosphatase (1-mole burst) is formed at low substrate concentrations, and a diphosphorylated derivative (2-mole burst) at high substrate concentrations (Figure 20). Again, in this case, in which a very low activity is observed, the available data appear to exclude the formation of the S—E—E—P complex.

There are three important events in the mechanism of the action of alkaline phosphatase: (1) binding of the substrate to the free enzyme or to the monophosphorylated derivative, (2) phosphorylation, and (3) dephosphorylation (Figure 17).

The considerable similarity of V_m and K_m values for very different substrates (Table 2) requires that these three events be independent of the chemical structure of the substrate. In fact, these predictions have been found to be fullfilled.

1. The free energy of binding is similar if not identical for most substrates, since the enzyme presents no specificity for the R part of $R-O-PO_3{}^{2-}$.

2. Dephosphorylation obviously occurs at the same rate with all substrates, whether or not the chemical transformation or a structural change is rate-determining.

3. The kinetic results obtained with Cu^{2+} phosphatase at alkaline

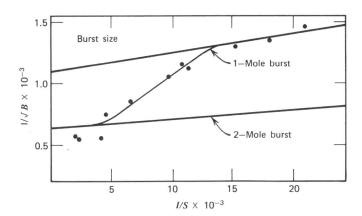

Figure 20 Burst amplitude (B) as a function of initial concentration of O-(p-phenylazo)phenylphosphorothioate by E. coli alkaline phosphatase. pH 8.5, 25°C. From reference 74.

pH (in this case phosphorylation is rate-limiting) have shown that the maximal rates are either identical or very similar for pyrophosphate and for several organic substrates with very different chemical structures. Therefore the phosphorylation step in the flip-flop mechanism cannot be controlled by the rate of the chemical reaction; it has to be controlled by the rate of structural change in the active center. This is the only way to have a phosphorylation process that occurs at a rate independent of the chemical nature of the leaving group [64].

Mechanisms of activity for *E. coli* phosphatase and intestinal alkaline phosphatase appear to be very similar. The latter seems to be an essential element of the phosphate transport system of the epithelial surface of the intestine [75]. It is strongly associated with the outer membranes of microvilli in the duodenum, jejunum, and ileum. The enzyme, dissociated from its membrane matrix, is a dimer of 140,000 molecular weight comprising two apparently identical subunits [76]. This glycoprotein contains four zinc gram atoms per mole of enzyme; it has two active sites and displays practically no specificity. Its catalytic activity is about 10 times that of *E. coli* phosphatase at alkaline pH (pH 8.5), and it is not very much influenced by the substrate structure.

That the two sites are not independent is demonstrated by the strong negative cooperativity for the binding of inorganic phosphate at pH 8.0 (Figure 21). The ratio of the dissociation constants observed for the phosphatase-(orthophosphate)$_1$ complex (K_1 = 4 μM) and the phosphatase-(orthophosphate)$_2$ complex (K_2 = 130 μM) indicates a free energy of interaction of 2.2 kcal mole^{-1} [77].

3 SUBUNIT INTERACTIONS IN DEHYDROGENASES

Extensive work has been carried out recently to show that subunits do not act independently in alcohol dehydrogenase.

3.1 Liver Alcohol Dehydrogenase

Liver alcohol dehydrogenase (LADH) is a dimer (mol wt 84,000) formed of identical subunits, each comprising 374 amino acids [78]. The dehydrogenase has two coenzyme binding sites [79–81] and contains four gram atoms of zinc per mole of protein [82]. The steady-state characteristics of the enzyme have been extensively studied, and it is well established that it obeys Michaelis-Menten

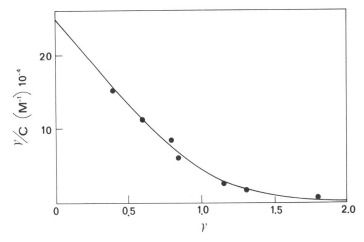

Figure 21 Scatchard plot for the binding of inorganic phosphate to calf intestine alkaline phosphatase. v, Mole binding ratio orthophosphate/dimer; c, concentration of free phosphate. pH 8.0, 25°C. 4 M NaCl. The solid curve is calculated for two sites with apparent dissociation constants $K_1 = 4\ \mu M$ and $K_2 = 130\ \mu M$.

kinetics [81]. LADH possesses a very broad specificity for its alcohol, aldehyde, and ketone substrates; the enzyme reacts with a variety of aromatic and aliphatic alcohols and a variety of aromatic and aliphatic aldehydes at similar maximal rates.

$$RCHO + NADH + H^+ \rightleftharpoons RCH_2OH + NAD^+$$

Subunit interactions for this enzyme are suggested by ^{35}Cl nuclear magnetic resonance (nmr) studies [83]. The chloride ion was used as a reporter group for conformational changes in horse LADH. When the enzyme was added to a KCl solution, the ^{35}Cl nmr signal became broadened as a result of interactions between the chloride ions and the dehydrogenase. After addition of 1 mole of NADH per mole of alcohol dehydrogenase, the ^{35}Cl nmr line width decreased to a level that remained unchanged upon addition of excess NADH.

The observed stoichiometry indicates that the initial identity of subunits in the coenzyme-free dimeric molecule disappears upon addition of 1 mole of coenzyme. The data show that binding of a single NADH molecule to the dimer is sufficient to affect chloride ion binding on both subunits, and strongly indicate an interaction between the two chains. The affinity of the second subunit for

NADH apparently decreases via intersubunit interactions.

The steady-state mechanism of ethanol-acetaldehyde catalysis is compulsory-ordered (the coenzyme molecule has to bind first), and coenzyme dissociation is usually believed to be rate-limiting [84, 85], as shown in the following scheme.

$$E + NAD^+ \rightleftharpoons E(NAD^+) \xrightarrow{C_2H_5OH} E(NAD^+) (C_2H_5OH) \rightleftharpoons$$

$$E(NADH) (CH_3CHO) \rightleftharpoons E(NADH) + CH_3CHO$$

Recent studies of transients in the reduction of aromatic aldehydes showed the kinetic nonequivalence of alcohol dehydrogenase subunits and cooperative subunit interaction in the ternary complexes of the enzyme [86–89]. The particular aldehydes chosen in these studies have spectral properties that readily allow the observation of either the aldehyde → alcohol or the NADH → NAD$^+$ transformation during catalysis. For example, benzaldehyde (1) allows the unperturbed observation of the NADH spectral change, whereas the long-wavelength π,π^* transition of the azoaldehyde, 4-(2′-imidazolylazo)benzaldehyde (2) allows spectral measurements at wavelengths remote from the enzyme and the NADH absorption bands.

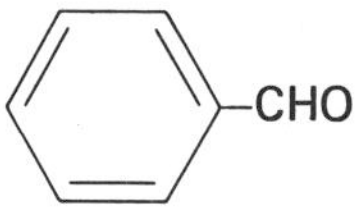

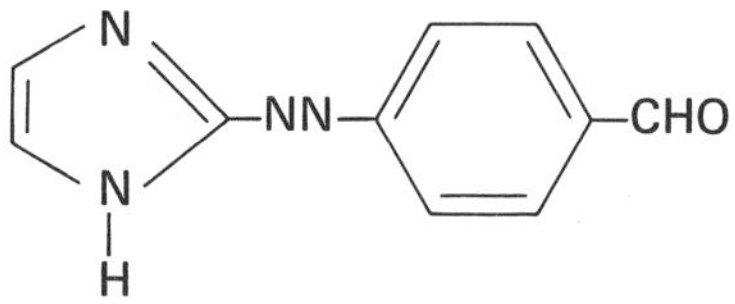

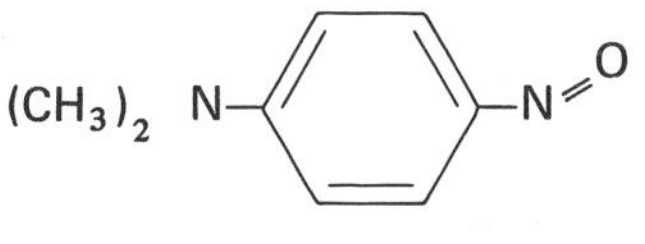

Benzaldehyde, azoaldehyde, and acetaldehyde reductions proceed with similar maximal rates [86–88]. An interesting study was also carried out with the substrate analog p-nitroso-N,N-dimethylaniline (NDMA) (3). Considerable spectral changes occur at 440 nm upon LADH-catalyzed reduction of (3). The excellent spectral properties of this system allow the independent observation of NADH at 330 nm and of (3) at 440 nm during reaction [87].

The stopped-flow analysis was carried out under conditions of limiting enzyme concentration ([E] < [NADH], [aldehyde]) and under conditions of limiting substrate concentration ([NADH] > [E] > [aldehyde]).

The conversion of reactants into products occurs via two distinct kinetic processes when the enzyme concentration exceeds that of NADH (Figure 22). The two first-order steps, one fast and one slower, involve transformations of equal amplitude.

When the amplitude of the rapid initial step is measured as a function of NADH concentration and plotted as in Figure 23, a titration curve giving the stoichiometry of the reacting sites is obtained. The ordinate intercept at the titration end point (P_{max}) is a measure of the stoichiometric relationship between the moles of reactants consumed in the burst and the enzyme-site concentration. The abscissa intercept (ϕ) is a measure of the stoichiometric

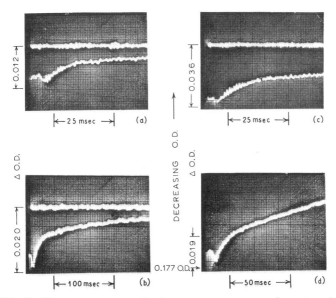

Figure 22 Oscilloscope traces of the progress curves for the change in transmission at 440 nm as a function of NADH concentration for the LADH-catalyzed reduction of NDMA by NADH. Conditions: $[NDMA]_0$ = 4.97 μM; $[LADH]_0$ = 1.06 μM; $[NADH]_0$ trace (a), 0.34 μM; (b) 0.564 μM; (c) 1.04 μM; and (d) 4.6 μM. pH 8.75 at 25°C. K_m values are 0.7 μM for NDMA and 2 μM for NADH. From reference 87.

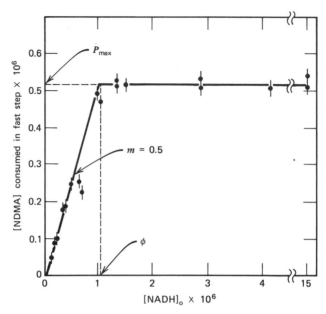

Figure 23   Dependence of the amount (moles per liter) of NDMA consumed in the rapid initial step (see Fig. 22) as a function of NADH concentration. Conditions: $[NDMA]_0$, 5 μM; $[LADH]_0$, 1.06 μM; pH 8.75 at 25°C. The initial solid line is theoretical for a slope (m) of 0.5. The ordinate and abscissa intercepts of this line with the horizontal end-point line, P_{max} and ϕ, respectively, are indicated. From reference 87.

relationship between the equivalents of enzyme sites required for the burst and the equivalents of enzyme-coenzyme binding sites. Figure 23 indicates an initial slope (m) of 0.5 mole of 3 consumed per mole of NADH, a value of 0.5 mole for the number of equivalents of 3 consumed per NADH-saturated site (P_{max}/N) and a 1:1 correspondence of ϕ and enzyme normality N. These findings parallel those described for several aldehyde substrates [86].

Half of the chemical transformation occurring at a slower rate than the other half is strongly suggestive of a cooperative and ordered function of two sites per dimeric enzyme.

This conclusion has been recently confirmed by McFarland and Bernhard, using a method termed "catalytic-site suicide" [89]. This technique permits analysis of the single turnover reduction of aldehyde substrates under conditions of large excess of both substrate and coenzyme over enzyme sites. The procedure involves poisoning of the catalytic reaction with pyrazole. Pyrazole reacts with the enzyme-NAD$^+$ complex; the bimolecular addition to E(NAD$^+$) can be made very rapid with high concentrations of pyrazole, and the dissociation of the ternary complex is very slow. Pyrazole poisoning of the active site during aldehyde reduction by NADH occurs only and immediately after alcohol product desorption:

$$E(NADH) \ (RCHO) \ (H^+) \ \rightleftharpoons \ E(NAD^+) \ (RCH_2OH) \ \rightleftharpoons \ E(NAD^+)$$

$$+ \ RCH_2OH \ \rightleftharpoons \ E(NAD^+) \ (pyrazole)$$

The turnover number per site is 1×10^{-3} sec^{-1} in the presence of pyrazole, as compared to 4.0 sec^{-1} in its absence. The former rate constant is presumably the specific rate of dissociation of the ternary complex E(NAD$^+$)(pyrazole) [89]. Therefore, the ternary complex, once formed, is sufficiently stable to prevent turnover during the time course of rapid kinetic observation. Each site in the dimeric alcohol dehydrogenase is "inactivated" by pyrazole after a single turnover reduction.

Here again a biphasic rate behavior for the formation of products, without detectable intermediates, was observed. The reduction of aromatic aldehydes proceeds via two kinetically distinct steps under a variety of conditions. Rapid kinetic studies with 4-deuterio-NADH have demonstrated a deuterium isotope effect of about 2.5 on the fast step, but no such deuterium isotope effect on the slow step. The most probable interpretation of these results is that the first site in the dimeric alcohol dehydrogenase reacts rapidly at a rate that is at least partially controlled by the chemical transfer of hydrogen from coenzyme to product (a primary isotope effect is observed only in rapid transients), whereas reaction at the second site must await isomerization of the enzyme triggered by desorption of the alcohol product from the first site [89].

The postulated asymmetry of the two subunits after binding of the coenzyme is consistent with the x-ray observations of Bränden and his group [2]. These workers observed that in the absence of coenzyme the dimeric alcohol dehydrogenase has a twofold axis of symmetry. This symmetry is lost in crystals of the enzyme containing bound coenzyme. Indications that the coenzyme produces an activation of one of the two enzyme sites, probably via a slow isomerization of the ternary complex, have been recently obtained by Luisi and Favilla [88].

3.2 Malate Dehydrogenase

Malate dehydrogenase (MDH) is another dimeric enzyme. It catalyzes the following reaction.

$$NAD^+ + \text{L-malate} \rightleftharpoons \text{oxaloacetate} + NADH + H^+$$

There are two types of MDH in the cardiac muscle cell: the mitochondrial (m-MDH) and the soluble, cytoplasmic (s-MDH) species. Both enzymes are dimers with similar molecular weights of 70,000–74,000 [90, 91]. Studies on the primary and three-dimensional structure of s-MDH by sequence and x-ray crystallographic methods has produced a 3.0-Å model [3].

The dimeric structure is most probably composed of identical subunits [92, 94]. m-MDH and s-MDH catalyze the same reaction and present Michaelian kinetics for coenzyme and substrates, malate and oxaloacetate [3, 95]. As for alcohol dehydrogenase, the mechanism of action of MDH is compulsory ordered; coenzyme has to bind first to "prepare" the subunit for oxaloacetate or L-malate binding (intrasubunit cooperativity) [3, 96–98].

Recent work carried out on s-MDH by Banaszak and his co-workers showed that the principal crystal form (type-C crystals) of the enzyme used in their x-ray crystallographic studies contained a single NAD^+ per dimeric molecule. Moreover, reaction of these crystals with p-hydroxymercuriphenyl sulfonate (PHMS) indicates the presence of three mercury sites per dimer. The most simple interpretation of these data is that the subunits are not independent in the dimer, and that negative cooperativity is very strong for NAD^+ binding. The coenzyme-free enzyme is thought to be a symmetric dimer [3]. Association of NAD^+ with one subunit induces a conformational change in the other subunit, which prevents binding of a second NAD^+ molecule and unmasks a sulfhydryl group to produce a third PHMS site. The asymmetry provoked by NAD^+ binding to a single subunit is probably very limited, since most regions on the electron density map of type-C crystals are extremely similar in pairs and apparently related by a local dyad [3, 96]. When type-C crystals were soaked in 20 mM NAD^+ plus 20 mM L-malate, or in 20 mM NAD^+ plus 20 mM oxaloacetate, there appeared, in addition to the NAD^+_p site (p for primary) a second binding site for NAD^+ which was designated the NAD^+_s site (s for secondary). Chemical studies indicate that this site is intimately associated with the catalytic mechanism [3].

All these data taken together show that subunits are not independent in MDH, although both s-MDH and m-MDH present Michaelis-Menten kinetics. This strongly suggests a coupled function of the two subunits in catalysis. A "reciprocating compulsory order mechanism" has already been proposed by Harada and Wolfe for m-MDH from a detailed analysis of steady-state kinetics [97, 98]. This mechanism suggests the intermediate formation of a complex

$$
\begin{array}{c}
\text{E--NAD, M} \\
| \\
\text{E--NADH, O}
\end{array}
$$

which contains 1 mole of NAD$^+$ and 1 mole of malate, **M**, on one subunit, and 1 mole of NADH and 1 mole of oxaloacetate, **O**, on the other one.

3.3 Glyceraldehyde-3-Phosphate Dehydrogenase

Glyceraldehyde-3-phosphate dehydrogenase (GAPDH) is an important enzyme in intermediary metabolism. It catalyzes the oxidation and subsequent phosphorylation of aldehyde substrates to their corresponding acyl phosphates: Glyceraldehyde-3-phosphate + NAD$^+$ + P$_i$ $\rightleftharpoons$ 1,3-diphosphoglycerate + NADH + H$^+$.

The oxidative phosphorylation of D-glyceraldehyde-3-phosphate takes place in two steps

$$
\text{RCHO} + \text{NAD}^+ + \text{E--SH} \longrightarrow \text{R}-\overset{\overset{\displaystyle O}{\|}}{\text{C}}-\text{S--E} + \text{NADH} + \text{H}^+
$$

$$
\text{R}-\overset{\overset{\displaystyle O}{\|}}{\text{C}}-\text{S--E} + \text{P}_i \rightleftharpoons \text{R}-\overset{\overset{\displaystyle O}{\|}}{\text{C}}-\text{OPO}_3{}^{2-} + \text{E--SH}
$$

In the first step the aldehyde reacts with the enzyme active site to form a thiol ester intermediate. The 3-phosphoglyceroyl enzyme is deacylated in the second step to produce the acyl phosphate.

The enzyme has been prepared from a variety of sources including microorganisms, fishes, birds, and mammals. It is always found as a tetramer of approximately 150,000 daltons. In yeast, lobster, and pig enzymes, the four subunits have been shown to be chemically identical. Each subunit of the tetramer binds one molecule of coenzyme and has one cysteine residue essential for enzymic activity.

A considerable amount of work has been done with muscle GAPDH. The enzyme presents classic steady-state kinetics with all its substrates (including both oxidation forms of the coenzyme) [99], but the active sites on each subunit are not independent. Both NAD⁺ and NADH binding show negative cooperativity [100–102]. As could be expected from its tetrameric structure, the enzyme binds four molecules of coenzyme. However, the binding constants differ greatly. The first two molecules of coenzyme are bound very firmly $(K_D < 0.5 \ \mu M)$, too firmly in fact to allow a determination of the dissociation constants. The third molecule is bound less tightly than the first two $(K_D = 2-5 \ \mu M)$, and the fourth molecule still less tightly $(K_D = 26-50 \ \mu M)$. While coenzyme binding data clearly imply the existence of intersubunit interactions, they do not clearly show that GAPDH is a half-site enzyme. An indication that this is so is to be found in the results of Givol [103], who observed that an active-site-directed reagent, p,p'-difluoro-m,m'-dinitrophenylsulfone, binds covalently to only two of the four subunits. GAPDH might then possibly be a double dimer $(\alpha\alpha)_2$. This conclusion corroborates the independent evidence obtained by using an active-site-specific acylating reagent [104, 105].

Since a thiol ester acyl enzyme is an essential intermediate in the catalytic mechanism, Bernhard and his group [104, 105] studied the effect of covalent thiol modification on the properties of rabbit muscle GAPDH. The tetrameric enzyme contains four particularly reactive thiols. By chemical modifications it was shown that they react with substrates to form the intermediate thiol ester derivative [104–106]. There is no preferential alkylation by iodoacetate or iodoacetamide of any of the four essential thiols [104, 105, 107]. Four moles of alkyl groups are required to inactivate the enzyme, and the percentage of inhibition is directly proportional to the degree of alkylation. However the four active-site thiols that appear to be identical and independent for "nonspecific" thiol alkylating agents do not appear to be identical and independent in catalytic reactivity. The 3-phosphoglyceroyl-enzyme can be formed readily by incubation of GAPDH with 3-phosphoglyceroyl phosphate. The 3-phosphoglyceroyl-enzyme is a true enzyme-substrate complex, and the fact that it is fairly stable suggests that this derivative accumulates under steady-state conditions (deacylation would then be the rate-limiting step of the whole process). Only two of the four essential thiol groups of the tetrameric GAPDH are acylated in the 3-phosphoglyceroyl-enzyme [104]. More extensive studies, using the pseudo substrate β-(2-furyl)acryloyl phosphate as an acylating reagent also showed that the reaction is complete when only two acyl groups are incorporated per mole of tetrameric enzyme [104]

(Fig. 24). Moreover, in accordance with these results, Keleti [108] showed that there is a transient release of only 2 moles of NADH in the action of fully saturated GAPDH [E(NAD$^+$)$_4$] on glyceraldehyde 3-phosphate.

All these results show very clearly that GAPDH is a half-site enzyme. The functional unit of the tetrameric structure is the α_2 dimer. Analysis of the kinetics and of the stoichiometry of acylation by β-(2-furyl)acryloyl phosphate indicates that the two protomers (α_2 dimers) of GAPDH are identical and independent in the catalytic mechanism when the NAD$^+$ concentration is high enough to saturate the four coenzyme sites.

A recent study by MacQuarrie and Bernhard [105] provides other evidence for the need of a basic asymmetry for the catalytic function of rabbit muscle GAPDH. Using the fact that in the tetrameric enzyme only two of the four essential thiol groups are readily acylated in the 3-phosphoglyceroyl-enzyme, a diacylated dicarboxy-methylated enzyme was prepared by reaction of the acyl enzyme with iodoacetate. The two acyl groups were then removed in the presence of phosphate or arsenate, but only one could be reintroduced. This apparent "irreversibility" demands an intra-molecular subunit rearrangement of the tetrameric structure. Figure 25 proposes a mechanism to explain the stoichiometry of acylation of muscle GAPDH.

If the mechanism is correct, it is of interest to note not only that the enzyme behaves as a double dimer in catalysis, but also that it has a "built-in" capacity for preserving the property of asymmetry necessary for its catalytic action.

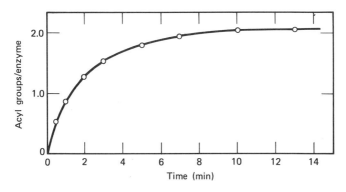

Figure 24   Moles of acyl groups per mole of GAPDH as a function of time at pH 7 and 25°C. [GAPDH] = 7.93 μM; [NAD$^+$] = 28 μM; [β-(2-furyl)acryloyl phosphate] = 6.9 10^{-4} M. From reference 104.

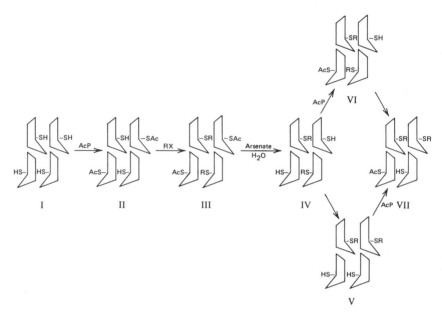

Figure 25 Possible mechanism to explain acylation stoichiometry of native [E–(SH)$_4$] and modified [(RS)$_2$-E-(SH$_2$)] GAPDH. AcP, Acyl phosphate; RX, iodoacetate. In such a mechanism species VII must be thermodynamically favored relative to species VI. From reference 105.

Recent crystallographic data have indicated that subunits are arranged with almost perfect 222 symmetry in coenzyme-free GAPDH from human muscle. However, local asymmetry has also been observed, suggesting that the active site might already be related in pairs in the holoenzyme [8]. Muscle GAPDH is obviously a good candidate for a flip-flop-type mechanism.

The relationship between the negatively cooperative binding of the coenzyme and its effect on the kinetics is difficult to appreciate. Although GAPDH from different sources behaves as half-site enzymes, the type of cooperativity for coenzyme binding differs considerably. Yeast GAPDH is also a tetramer, showing half-of-the-sites reactivity for acylation by β-(2-furyl)acryloyl phosphate [104]. However, in contrast to the GAPDH from most mammalian muscle sources and from lobster, it binds NAD$^+$ with much higher affinity and with a sigmoidal binding isotherm (positive cooperativity) at alkaline pH [109]. Coenzyme dissociation constants are such that the enzyme is fully or nearly fully saturated with NAD$^+$ [E(NAD$^+$)$_4$] under the usual conditions of steady-state kinetics.

3.4   Lactate Dehydrogenase

Lactate dehydrogenase (LDH) is a tetrameric enzyme of molecular weight 140,000 daltons, which catalyzes the following reaction:

$$\text{L-Lactate} + \text{NAD}^+ \rightleftharpoons \text{pyruvate} + \text{NADH} + \text{H}^+$$

The functional significance of the multisubunit structure is much less documented than for alcohol-, malate-, or GAPDH. However, the LDH mechanism, like that of alcohol- or MDHs, proceeds via an ordered binding of coenzyme followed by substrates [110, 111], and kinetics are of the Michaelis-Menten type [112]. Furthermore, the tertiary structure of the LDH subunit is extremely similar to that of the MDH subunit [113]. This resemblance suggests that subunit interactions, which are of importance for catalysis in other dehydrogenases, also exist in LDH.

The x-ray analysis of crystals of the M_4 isoenzyme (skeletal muscle) from dogfish has shown that the tetramer has a 222 symmetry in the absence of coenzyme [114]. When apoenzyme crystals are soaked in ammonium sulfate solutions containing 1.4 mM NAD$^+$ or 0.05 mM NADH, the coenzyme is bound with an accompanying change in the crystal lattice. This change indicates a new space group (change of the space group from F 422 to C 42 21 2) for the coenzyme containing dehydrogenase crystals in which the tetramer has only one twofold axis of symmetry. This reduction of molecular symmetry from 222 to 2 by diffusion of the coenzyme into single crystals of the dogfish LDH suggests that the enzyme, like GAPDH, behaves as a double dimer in catalysis. Crystals of the abortive ternary complex of dogfish LDH, NAD$^+$ (3.9 mM), and pyruvate (0.1 M) have been obtained in ammonium sulfate solution [114]. The tetrameric abortive complex again has 222 symmetry, although the packing of molecules with respect to the tetragonal axis is different from that in the apoenzyme crystals.

3.5   L-Glutamate Dehydrogenase

L-Glutamate dehydrogenase (GDH) is located at the intersection of several metabolic pathways. It catalyzes the following reaction.

$$\text{L-Glutamate} + \text{NAD(P)}^+ + \text{H}_2\text{O} \rightleftharpoons \alpha\text{-ketoglutarate} + \text{NAD(P)H} + \text{NH}_4^+$$

The enzyme is arranged as a hexamer composed of identical subunits, and the molecular weight of the subunits is 52,000 daltons [115]. One active site per subunit was found for the coenzyme and for α-ketoglutarate [116, 117]. Shafer and co-workers measured by the stopped-flow technique the amount of NADH [118] formed in the initial rapid phase of the presteady-state phase under conditions where [NAD$^+$] > [GDH]. A study of the dependence of the presteady-state kinetics on NAD$^+$ concentration revealed that, between 27 and 1000 μM NAD$^+$ the stoichiometry was invariant at about 0.4 moles of NADH formed per subunit in the burst phase, that is, about 3 moles of NADH per hexamer. This observation suggests that half-site reactivity also operates in this dehydrogenase.

4 SOME PHENOMENOLOGICAL ASPECTS OF FLIP-FLOP MECHANISMS

In addition to phosphatases and dehydrogenases, an increasing number of Michaelian enzymes appears to present absolute negative cooperativity for substrate binding and/or apparent reactivity of half of the sites in the catalytic reaction (Table 4). The latter indicates that the subunits are not always independent in catalysis.

It is easy to conceive of a variety of mechanisms in which there is a functional coupling of the two subunits of a dimer [126]. These mechanisms generally give a v versus S profile of the Michaelian type. For example, one can write 32 different mechanisms for an enzyme made up of functional dimers that use two substrates and two products (without formation of a covalent intermediate), depending on the type of coupling: binding of the substrate to one subunit and chemical reaction on the other one; binding of the substrate and product exclusion; and so on. Only some typical properties of these mechanisms are discussed here.

A typical mechanism is presented in Figure 26. In such a mechanism two steps appear only in the transient phase, and they are not seen under steady-state conditions. The introduction of S_1 and S_2 only in the sequential steps implies that the mechanism is of the Michaelian type.

Another mechanism, presented in Figure 27, indicates that a catalytic mechanism can very well implicate a coupled function of the subunits and still obey (1) steady-state kinetics of the Michaelian type, (2) first-order transient kinetics, and (3) formation of products in the presteady-state phase corresponding to 2 moles of P_1 or P_2 liberated per mole of dimer. Such properties are generally

TABLE 4 SOME ENZYMES WITH HALF-OF-THE-SITES REACTIVITY

Enzyme	Number of Subunits	Number of Functional Sites	Substrate or Analog Tested	Reference
m-Malate dehydrogenase	2	1	NAD⁺	3, 96
Alcohol dehydrogenase	2	1	NADH, aldehydes, p-nitroso-N, N-dimethylaniline	86, 89
Transaldolase	2	1	Dihydroxyacetone	119, 120
Alkaline phosphatase (E. coli)	2	1	P$_i$, arsenate, organic phosphates, pyrophosphate	32, 61–64
Muscle glyceraldehyde-3-phosphate dehydrogenase	4	2	3-phosphoglyceroyl phosphate, β-(2-furyl)acryloyl phosphate	104 105
CTP synthetase	4	2	6-diazo-5-oxonorleucine	122
Methionine tRNA synthetase (E. coli)	4	2	L-Methionine L-Methioninyl adenylate	121
Glutamate dehydrogenase	6	3	NAD⁺	118
Glutamine synthetase	8	4	Methionine sulfoximine	123
Acetoacetate decarboxylase	12	6	Acetic anhydride, 2, 4 dinitro-phenyl acetate, acetopyruvate	124, 125

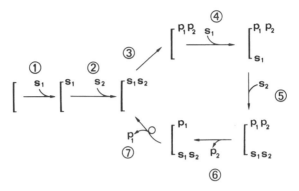

Figure 26 A typical flip-flop-type mechanism for a functional dimer with two substrates and two products. In that particular case intersubunit cooperativity couples: (1) product formation on the first subunit (step 3) and binding of S_1 to the second one (step 4); (2) binding of S_2 to the second subunit (step 5) and release of P_1, then of P_2 from the first one (steps 6 and 7). From reference 126.

believed to characterize independent subunits. In this mechanism for a functional dimer with two substrates and two products, binding of S_1 is anticooperative, and binding of S_2 is coupled (intrasubunit coupling) to S_1 binding on the same subunit. When S_1 and S_2 are bound to the subunit, they can react and form P_1 and P_2 (step 3). This step is the initiation of the steady-state cycle. Steps 1, 2, or 3 are found only in the presteady-state phase. In such a mechanism, a complex

$$\begin{bmatrix} P_1 P_2 \\ P_1 P_2 \end{bmatrix}$$

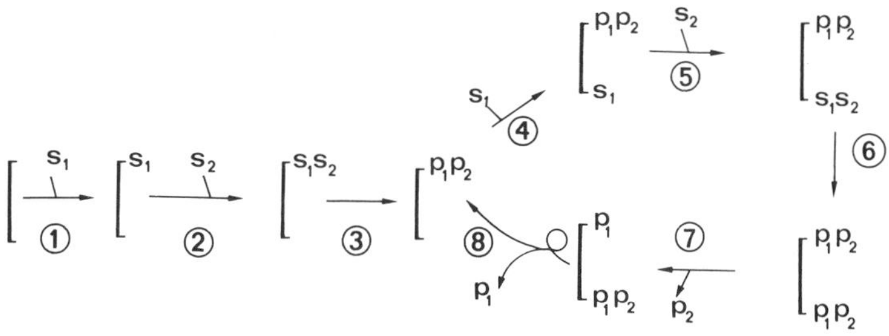

Figure 27 A flip-flop-type mechanism with intermediate formation of a $\begin{bmatrix} P_1\ P_2 \\ P_1\ P_2 \end{bmatrix}$ complex. From reference 126.

would be formed after the two subunits have reacted. In step 7, the chemical reaction on the second subunit is coupled with the liberation of P_2 from the first one. P_1 is eliminated in step 8. The looped arrow indicates that we then have a complex with a single molecule of P_1 and a single molecule of P_2. Another turn in the cycle is necessary to eliminate products from the second active site. Each steady-state cycle requires the introduction of a single molecule of S_1 and of a single molecule of S_2; such a mechanism then gives Michaelian kinetics for S_1 and S_2. The complex that accumulates under steady-state conditions is

$$\begin{bmatrix} P_1 P_2 \\ P_1 P_2 \end{bmatrix}$$

when step 7 is rate-limiting. Under these conditions there are 2 moles of P_1 and of P_2 produced per mole of dimer at the end of the presteady-state period (burst). The transient phase is first-order when step 3 is limiting.

With the same mechanism and a substrate S_1 or S_2 with a different structure, step 5 could replace step 7 as the slowest step of the steady state. In such a case the presteady-state burst of P_1 and P_2 would produce only 1 mole per mole of enzyme.

A mechanism of this type (not necessarily identical) could explain the apparently paradoxical situation that exists for alcohol dehydrogenase. Shore and Gutfreund [127], using acetaldehyde as substrate, found a burst of 2 moles of NAD^+ per mole of enzyme, whereas Bernhard, Luisi, and their collaborators [86–88], using aromatic aldehydes, found a burst of only 1 mole of NAD^+ per mole of dimeric enzyme in the presteady-state period. There is little doubt that the basic mechanism of action of alcohol dehydrogenase is the same for both types of substrates. The steady-state cycle in Figure 27 is effectively the same for the two kinds of substrates; but step 4, 5, or 6 would be limiting for substrates of the benzaldehyde type, whereas step 7 or 8 could be limiting for substrates like acetaldehyde.

Although in Figures 26 and 27 negative cooperativity is observed for all ligands, this is not a necessary condition for a flip-flop-type mechanism. In Figure 28 is a mechanism in which negative cooperativity exists for S_2 but not for S_1. In most cases, when there is an ordered binding of substrates, equilibrium dialysis is carried out with the substrate that binds first. In such a case equilibrium dialysis with S_1 suggests an independence of the sites.

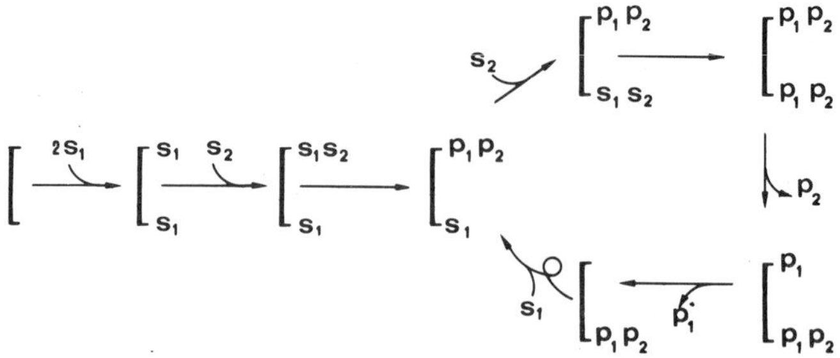

Figure 28   A typical mechanism involving negative cooperativity for S_2 but not for S_1. From reference 126.

The preceding schemes did not take into consideration the influence of products on the reaction. Figure 29 is a typical scheme demonstrating the influence of the products on flip-flop-type kinetics.

The presence of products can have various effects.

If steps $1'$ and $2'$, outside the steady-state cycle, are more rapid than steps 1 and 2, the presteady-state phase of the reaction in the direction $S_1 + S_2 \rightarrow P_1 + P_2$ may be accelerated in the presence of products.

Finally, it is worthwhile to point out that negative cooperativity may be obtained for the binding of the product and not necessarily for the binding of the corresponding substrate. In Figure 30 it is shown that, although the binding of P_1 is anticooperative, this is not

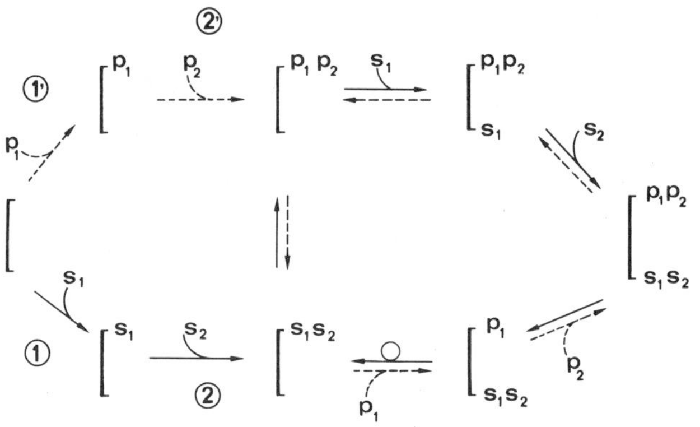

Figure 29   The influence of products on flip-flop-type kinetics. From reference 126.

true of the binding of S_1. Kinetics are of the Michaelian type for S_1, S_2, P_1, and P_2. In the particular case of Figure 30, the presence of S_1 (step $3'$) is necessary to start the reaction in the direction $P_1 + P_2 \rightarrow S_1 + S_2$.

We find here observations that have been made several times. For example, negative cooperativity has been observed in NAD^+ binding to MDH [3] as previously described, but not in NADH binding. NADH binding sites appear to be identical and independent [128]. However, an absolute requirement for NAD^+ has been observed for the NADH oxidation by acetylphosphate catalyzed by muscle GAPDH [129].

Flip-flop-type mechanisms can also explain activation of inhibition caused by excess of substrate, which may be of interest in the autoregulation of the catalytic reaction. Figure 31 presents a very simple example to explain the kinetic effects of an excess of substrate. Many other such mechanisms could be written.

Cycle I describes a classic flip-flop-type mechanism. We suppose that the rate-limiting step in this cycle is step 5, that is, the intersubunit conformational change that couples binding of substrate S_1 to the second subunit with exclusion of P_2 from the first one. The characteristic kinetic properties of cycle I are those of a Michaelian enzyme. This cycle can be branched with cycle II. Cycle

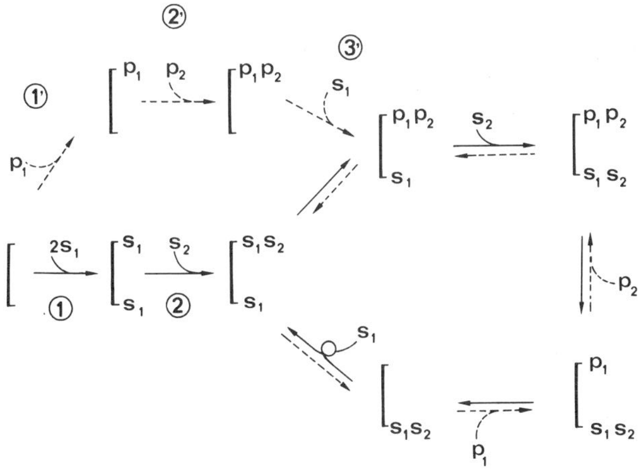

Figure 30 A flip-flop-type mechanism with different binding properties for substrate S_1 (independent binding sites) and its corresponding product P_1 (negative cooperativity). From reference 126.

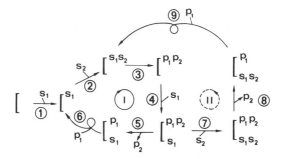

Figure 31   Inhibition or activation by excess of substrate S_2. From reference 126.

II can be started only at high concentrations of substrate S_2, when step 7 becomes possible. Under such conditions, when step 8 is slower than step 5, inhibition is caused by excess of substrate S_2; when it is faster, activation occurs.

As is well known, there are numerous examples in the literature of inhibitions or activations caused by excess of substrate, and although there are various ways to explain such phenomena [31, 34, 130, 131], situations such as that described in Figure 31 appear to provide a very simple explanation.

The last problem to be examined in this article is that of the compatibility between sigmoidal kinetics of allosteric enzymes and the possible occurrence of flip-flop-type mechanisms in the action of these enzymes. Negative cooperativity or half-of-the-sites reactivity is a characteristic property of flip-flop mechanisms. Although the characteristic feature of allosteric kinetics is positive cooperativity, this does not exclude a flip-flop mechanism as a basic catalytic process.

For the sake of simplicity, our model of an allosteric enzyme is a tetramer, as described by Monod, Wyman, and Changeux [22], which undergoes a concerted transition from the T to the R state. Only substrate binding will be discussed.

$$R \rightleftharpoons T$$

If the four homologous catalytic sites are intrinsically independent, the saturation function will be [22]:

$$\overline{Y_S} = \frac{Lc\alpha(1 + c\alpha)^3 + \alpha(1 + \alpha)^3}{L(1 + c\alpha)^4 + (1 + \alpha)^4}$$

where $L = T/R$ (in the absence of substrate), $\alpha = S/K_R$, and $c = K_R/K_T$. K_R and K_T are the microscopic constants for the dissociation of the substrate from its specific binding site in the R and T state, respectively. This equation for an equilibrium situation is also valuable for steady-state conditions in which dissociation constants K_R and K_T are replaced by corresponding K_m values [132]. Such a model with four independent catalytic sites is not of the flip-flop type. The maximal value of the Hill coefficient obtained from v versus S profiles will be near $n = 4$ for $c = 0$ (or very near 0), and for high values of L (over 10^4) [134].

Let us suppose now that this tetramer is a double dimer. The functional unit in the new model is the dimeric structure; dimers work catalytically as identical and independent flip-flop units, and each functional dimer displays half-of-the-sites reactivity for the substrate. In this case absolute negative cooperativity characterizes the functional dimer, whereas positive cooperativity characterizes interdimer subunit interactions. The saturation function is:

$$\overline{Y_S} = \frac{Lc\alpha(1 + c\alpha) + \alpha(1 + c\alpha)}{L(1 + c\alpha)^2 + (1 + \alpha)^2}$$

Preferential binding of the substrate to the R state obviously imposes a sigmoidal shape on the v versus S profile. Binding and catalysis exclusively by the R state and a very large L value result in a Hill coefficient of 2 or nearly 2, instead of nearly 4 as in the previous model. The "half-site allosteric enzyme" binds a maximum number of two substrate molecules on its four subunits (anticooperativity), and the binding isotherm indicates a maximum value of 2 for the Hill number, that is, half the number of subunits.

Half-of-the-sites reactivity in an allosteric enzyme similar to our model will be manifested not only by the number of substrate binding sites (two instead of four, for example) but also by the low value of the Hill coefficients.

Several allosteric enzymes were found to have Hill coefficients practically equal to the number of subunits. For example, Hill numbers of 1.8 to 2.0 have been found for saturation of the dimeric phosphorylase b by the effector AMP [133]. Therefore, while several interpretations may be given to Hill coefficients lower than the maximal value [22, 26, 34, 134] (and references therein), we feel that Hill coefficients (obtained from substrate binding or catalysis) corresponding to half or less than half the maximal value (i.e., the number of identical subunits) could be indications of a flip-flop-type mechanism.

A double dimer with two substrates may well have different Hill coefficients for each of its substrates. For example independent binding of substrate S_1 to each one of the four subunits, and negative cooperativity for substrate S_2 (only two of the four active sites can be saturated) may result in a flip-flop mechanism. If binding and catalysis are carried out exclusively by the R state (L very high) the maximal values of the Hill coefficients will be 4 for substrate S_1, and only 2 for substrate S_2.

If binding S_1 is carried out exclusively by the R state with no preferential anticooperative binding of S_2 to R or T, catalysis will occur only in the R state, the maximal value of the Hill coefficient will be 4 for S_1, and a Langmuir-type profile will be observed with S_2.

The allosteric systems just discussed are "K systems" [22] as defined by Monod et al. A flip-flop mechanism with an allosteric dimer gives a "V system" [22]. The dimer presents negative cooperativity for the substrate and positive cooperativity only for the other effectors (activators or inhibitors not structurally related to the substrate). Langmuir-type profiles are always observed for saturation by the substrate, and sigmoidal profiles are obtained with effectors. For example, if L is very high, exclusive binding of the substrate to the R state, which works catalytically as a flip-flop unit (half-of-the-sites reactivity), will produce a Michaelis-Menten profile for substrate saturation. Exclusive binding of an inhibitor to both independent inhibitor sites in the T state gives an hyperbolic saturation function in the absence of substrate, and a sigmoidal saturation function in saturating concentrations of the substrate. Under the latter conditions, the Hill coefficient will be close to 2.

Allosteric enzymes presenting positive homotropic effects for some of their substrates and negative homotropic effects for others are also known. One of the best-studied examples is cytidine triphosphate (CTP) synthetase. CTP synthetase catalyzes the formation of CTP from NH_3, UTP, and ATP.

$$UTP + NH_3 + ATP \longrightarrow CTP + ADP + P_i$$

The enzyme can also utilize glutamine instead of NH_3 as a nitrogen donor when GTP is present as an allosteric effector [135]

$$UTP + Glu-NH_2 + ATP \xrightarrow{GTP} CTP + Glu-OH + ADP + P_i$$

The active enzyme is a tetramer with positive cooperative effects for the two tetramer-stabilizing substrates ATP and UTP, and negative cooperative effects for glutamine. Half-of-the-sites reactivity was demonstrated with an affinity label, 6-diazo-5-oxonorleucine, which mimics glutamine. Only 2 moles of 6-diazo-5-oxonorleucine have been found to bind covalently per mole of tetramer [122], indicating that the tetrameric CTP synthetase is a double dimer. Each functional dimer behaves as a half-site enzyme, and is a good candidate for a flip-flop-type mechanism. Absolute negative cooperativity for only one substrate suffices to impose catalytic asymmetry.

5 CONCLUSIONS

This chapter has attempted to summarize present information suggesting that intersubunit cooperativity is not restricted to allosteric enzymes. The alternating operation of the two subunits within a functional dimer might be the basic mode of action of many oligomeric enzymes. In flip-flop-type mechanisms, conformational changes are not mainly used for a modulation of the enzymic activity (as for the allosteric transition) [22]; they are an integral part of the catalytic mechanism itself. The importance of structural changes for the catalytic activity of enzymes was first formulated by Koshland in the form of the "induced-fit" theory [136].

What are the advantages of flip-flop-type mechanisms?

1. The alternation of conformations between the two subunits of a functional dimer ensures thermodynamic coupling between successive steps of the catalytic mechanism. In alkaline phosphatase, intrasubunit cooperativity couples substrate binding with phosphorylation; intersubunit cooperativity couples substrate binding (a second molecule) to dephosphorylation with product release.

2. Flip-flop-type mechanisms provide an explanation for various kinds of inhibition or activation caused by excess of substrate or by reaction products. Such effects are often difficult to explain with classic theories for ordered mechanisms [98, 130, 131]. These inhibitions or activations might be of importance for the autonomous control of enzymic activity at each step in a metabolic sequence.

3. Monod et al. [22], in their classic paper on allosteric transitions, discussed in the following terms the evolutionary advantage of having symmetric oligomers: "The structural and functional effects of single mutations occurring in a symmetrical oligomer, or allowing its

formation, should be greatly amplified as compared with the effects of similar mutations in a monomer or in a non-symmetrical dimer. In other words, because of the inherent cooperativity of their structure, symmetrical oligomers should constitute particularly sensitive targets for molecular evolution, allowing much stronger selective pressures to operate in the random pursuit of functionally adequate structures." Obviously, this evolutionary advantage is important only if there are both structural and functional interdependences between the subunits in the oligomers. These interdependences have been well established for allosteric enzymes [22, 26]. It has been shown in this article that they also exist for other polymeric enzymes not endowed with specific regulatory functions, such as phosphatases or dehydrogenases.

A very important aspect that has not been treated in this chapter concerns enzyme-enzyme interactions in multienzyme complexes. It is clear that aggregates of enzymes, each contributing a specific catalytic site, must be able to profit from quaternary structure in the same general way as monofunctional oligomers. The interaction between subunits can lead to a mutual stabilization of active conformations, and to indirect and coordinated control of enzyme activity. This type of regulation is illustrated by aspartokinase (I)-homoserine dehydrogenase (I) from E. coli [137], in which both associated catalytic activities are inhibited by threonine in a cooperative fashion.

In addition to their heterologous composition, multienzyme complexes appear to differ from monofunctional enzymes in another important way; the different active sites of several complexes, such as α-ketoglutarate dehydrogenase or pyruvate dehydrogenase complexes, appear to be closely juxtaposed [138-139].

Some of the advantages of such a spatial organization for promoting the efficiency of catalysis are easy to imagine. Juxtaposition of active sites involved in the catalysis of sequential metabolic reactions would be expected to increase overall catalytic efficiency by reducing the diffusion path for products (which are substrates of subsequent reactions). Moreover, metabolic intermediates could be prevented from diffusing into the surrounding medium. This would eliminate undesirable branching of a particular metabolic pathway and could also protect labile intermediates from spontaneous degradation [138-139]. Thus formation of multienzyme complexes appears to be a very efficient means of concentrating and compartmentalizing different but metabolically related enzymes at the molecular level [140]. Moreover, the binding of the initial substrate could "activate" the catalytic sites by indirect means involving conformational changes. Some of the essential features of the catalytic mechanism of some of the simplest multienzyme complexes, for example, tryptophan synthetase, are already known [24, 141].

It is probable that future mechanistic investigations of enzyme aggregates will provide more information concerning the inherent advantages offered by subunit-subunit interactions than is presently available from studies with simpler enzymic systems.

ACKNOWLEDGMENTS

The author thanks the following agencies for support of the research discussed in this chapter: the Centre National de la Recherche Scientifique (ATP 2105), the Délégation Générale à la Recherche Scientifique et Technique (n° 72.7.0472), the Fondation pour la Recherche Médicale.

Thanks are also due to Dr. Chappelet and Mrs. Aubert for their help in the preparation of the manuscript.

REFERENCES

1. A. W. Hanson, M. L. Applebury, J. E. Coleman, and H. W. Wyckoff, *J. Biol. Chem.*, 245, 4975 (1970).

2. C. I. Branden, E. Zeppezauer, T. Boiwe, G. Söderlund, B. O. Söderberg, and B. Nordström, in *Pyridine Nucleotide-Dependent Dehydrogenases*, H. Sund, Ed., Springer-Verlag, Berlin, 1970, p. 129.

3. D. Tsernoglou, E. Hill, and L. J. Banaszak, *Cold Spring Harbor Symp. Quant. Biol.*, 36, 171 (1971).

4. D. W. Banner, A. C. Bloomer, G. A. Petsko, D. C. Phillips, and C. I. Pogson, *Cold Spring Harbor Symp. Quant. Biol.*, 36, 151 (1972).

5. J. W. Campbell, E. Duée, G. Hodgson, W. D. Mercer, D. K. Stammers, P. L. Wendell, H. Muirhead, and H. C. Watson, *Cold Spring Harbor Symp. Quant. Biol.*, 36, 165 (1972).

6. T. A. Steitz, *J. Mol. Biol.*, 61, 695 (1971).

7. M. J. Adams, D. J. Haas, B. A. Jeffery, A. McPherson, H. L. Mermall, M. G. Rossman, R. W. Schevitz, and A. J. Wonacott, *J. Mol. Biol.*, 41, 159 (1969).

8. H. C. Watson, E. Duée, and W. C. Mercer, *Nature*, 240, 130 (1972).

9. J. W. Campbell, G. I. Hodgson, and H. C. Watson, *Nature*, 240, 137 (1972).

10. P. A. M. Eagles, L. N. Johnson, M. A. Joynson, C. H. McMurray, and H. Gutfreund, *J. Mol. Biol.*, 45, 533 (1969).

11. E. Penhoet, M. Kochman, R. C. Valentine, and W. J. Rutter, *Biochemistry*, 6, 2940 (1967).
12. P. A. M. Eagles and L. N. Johnson, *J. Mol. Biol.*, 64, 693 (1972).
13. N. A. Kiselev, F. Y. Lerner, and N. B. Livanova, *J. Mol. Biol.*, 62, 537 (1971).
14. Y. Morino and E. E. Snell, *J. Biol. Chem.*, 242, 5591 (1967).
15. R. Josephs, *J. Mol. Biol.*, 55, 147 (1971).
16. D. C. Wiley and W. N. Lipscomb, *Nature*, 218, 1119 (1968).
17. R. C. Valentine, B. Shapiro, and E. R. Stadtman, *Biochemistry*, 7, 2143 (1968).
18. D. Eisenberg, E. G. Heidner, P. Goodkin, M. N. Dastoor, B. H. Weber, F. Wedler, and J. D. Bell, *Cold Spring Harbor Symp. Quant. Biol.*, 36, 291 (1972).
19. I. M. Klotz, N. R. Langerman, and D. W. Darnall, *Ann. Rev. Biochem.*, 39, 25 (1970).
20. R. H. Haschemeyer, *Advan. Enzymol.*, 33, 71 (1970).
21. N. M. Green, in *Protein-Protein Interactions*, R. Jaenicke and E. Helmreich, Eds., Springer Verlag, Berlin, 1972, p. 183.
22. J. Monod, J. Wyman, and J. P. Changeux, *J. Mol. Biol.*, 12, 88 (1965).
23. J. Monod, in *Symmetry and Function*, Nobel Symposium, A. Engström and Strandberg, Eds., Vol. II, Wiley, New York, 1969, p. 15.
24. K. Kirschner and R. Wiskocil, in *Protein-Protein Interactions*, R. Jaenicke and E. Helmreich, Eds., Springer-Verlag, Berlin, 1972, p. 245.
25. J. C. Gerhardt, *Current Topics Cell. Regulation*, 2, 275 (1970).
26. D. E. Koshland, Jr., G. Nemethy, and D. Filmer, *Biochemistry*, 5, 365 (1966).
27. M. F. Perutz, *Nature*, 228, 726 (1970).
28. D. E. Atkinson, in *The Enzymes*, P. D. Boyer, Ed., Vol. 1, 3rd ed., Academic Press, New York, 1970, p. 461.
29. K. Kirchner, *Current Topics Cell. Regulation*, 4, 167 (1971).
30. D. E. Koshland, Jr. and K. E. Neet, *Ann. Rev. Biochem.*, 37, 359 (1968).
31. D. E. Koshland, Jr., in *The Enzymes*, P. D. Boyer, Ed., Vol. 1, 3rd ed., Academic Press, New York, 1970, p. 341.
32. M. Lazdunski, C. Petitclerc, D. Chappelet, and C. Lazdunski, *Eur. J. Biochem.*, 20, 124 (1971).
33. A. Conway and D. E. Koshland, Jr., *Biochemistry*, 7, 4011 (1968).
34. D. E. Koshland, Jr., *Current Topics Cell. Regulation*, 1, 1 (1969).
35. F. Rothman and R. Byrne, *J. Mol. Biol.*, 6, 330 (1963).
36. M. J. Schlesinger, *Brookhaven Symp. Biol.*, 17, 66 (1964).
37. D. J. Plocke, C. Levinthal, and B. L. Vallee, *Biochemistry*, 1, 373 (1962).
38. C. Lazdunski, C. Petitclerc, and M. Lazdunski, *Eur. J. Biochem.*, 8, 510 (1969).
39. R. T. Simpson and B. L. Vallee, *Biochemistry*, 7, 4343 (1968).
40. R. T. Simpson and B. L. Vallee, *Ann. N.Y. Acad. Sci.*, 166, 670 (1969).
41. B. L. Vallee and D. D. Ulmer, *Ann. Rev. Biochem.*, 41, 91 (1972).

42. C. Lazdunski, C. Petitclerc, D. Chappelet, F. Leterrier, P. Douzou, and M. Lazdunski, *Biochem. Biophys. Res. Commun.*, 40, 589 (1970).
43. D. Chappelet, C. Lazdunski, C. Petitclerc, and M. Lazdunski, *Biochem. Biophys. Res. Commun.*, 40, 91 (1970).
44. C. Lazdunski, D. Chappelet, C. Petitclerc, F. Leterrier, P. Douzou, and M. Lazdunski, *Eur. J. Biochem.*, 17, 239 (1970).
45. M. L. Applebury and J. E. Coleman, *J. Biol. Chem.*, 244, 308 (1969).
46. M. I. Harris and J. E. Coleman, *J. Biol. Chem.*, 243, 5063 (1968).
47. J. H. Schwartz, *Proc. Nat. Acad. Sci. U.S.*, 49, 871 (1963).
48. L. Engstrøm, *Ark. Kemi*, 19, 129 (1962).
49. G. H. Tait and B. L. Vallee, *Proc. Nat. Acad. Sci. U.S.*, 56, 1247 (1966).
50. R. T. Simpson and B. L. Vallee, *Biochemistry*, 9, 953 (1970).
51. C. Lazdunski and M. Lazdunski, *Biochim. Biophys. Acta*, 114, 551 (1966).
52. C. Lazdunski and M. Lazdunski, *Eur. J. Biochem.*, 7, 294 (1969).
53. J. E. Coleman, *J. Biol. Chem.*, 242, 5212 (1967).
54. M. E. Riepe and J. W. Wang, *J. Biol. Chem.*, 243, 2779 (1968).
55. M. E. Fabry, S. H. Koenig, and W. E. Schillinger, *J. Biol. Chem.*, 245, 4256 (1970).
56. C. Lazdunski, C. Petitclerc, D. Chappelet, and M. Lazdunski, *Biochem. Biophys. Res. Commun.*, 37, 744 (1969).
57. D. Levine, T. W. Reid, and I. B. Wilson, *Biochemistry*, 8, 2374 (1969).
58. T. W. Reid, M. Lavlic, D. Sullivan, and I. B. Wilson, *Biochemistry*, 8, 3184 (1969).
59. M. L. Applebury and J. E. Coleman, *J. Biol. Chem.*, 245, 4969 (1970).
60. M. L. Applebury, B. P. Johnson, and J. E. Coleman, *J. Biol. Chem.*, 245, 4968 (1970).
61. C. Petitclerc, C. Lazdunski, D. Chappelet, A. Moulin, and M. Lazdunski, *Eur. J. Biochem.*, 14, 301 (1970).
62. H. N. Fernley and P. G. Walker, *Nature*, 212, 1435 (1966).
63. S. H. D. Ko and F. J. Kézdy, *J. Am. Chem. Soc.*, 89, 7139 (1967).
64. D. R. Trentham and H. Gutfreund, *Biochem. J.*, 106, 455 (1968).
65. A Torriani, *Biochim. Biophys. Acta*, 38, 460 (1960).
66. D. P. Fan, M. J. Schlesinger, A. M. Torriani, K. J. Barrett, and C. Levinthal, *J. Mol. Biol.*, 15, 32 (1966).
67. M. J. Schlesinger, *J. Biol. Chem.*, 242, 1604 (1967).
68. M. J. Schlesinger, J. A. Reynolds, and S. Schlesinger, *Ann. N.Y. Acad. Sci.*, 166, 368 (1969).
69. M. J. Schlesinger, *J. Biol. Chem.*, 240, 4293 (1965).
70. M. L. Applebury and J. E. Coleman, *J. Biol. Chem.*, 244, 709 (1969).
71. H. Csopak and K. E. Falk, *FEBS Letters*, 7, 147 (1970).
72. S. E. Halford, N. G. Bennett, D. R. Trentham, and H. Gutfreund, *Biochem. J.*, 114, 243 (1969).
73. S. E. Halford, *Biochem. J.*, 125, 319 (1971).
74. J. F. Chlebowski and J. E. Coleman, *J. Biol. Chem.*, 247, 6008 (1972).
75. F. Moog and M. S. Glazier, *Comp. Biochem. Phys.*, 42, 321 (1972).

76. M. Fosset, D. Chappelet, and M. Lazdunski, *Biochemistry*, 13 (1974).
77. D. Chappelet, M. Fosset, M. Iwatsubo, C. Gache, and M. Lazdunski, *Biochemistry*, 13, 1788, (1974).
78. H. Jornvall, *Eur. J. Biochem.*, 16, 25 (1970).
79. S. R. Anderson and G. Weber, *Biochemistry*, 4, 1948 (1965).
80. G. Pfleiderer and F. Auricchio, *Biochem. Biophys. Res. Commun.*, 16, 53 (1964).
81. H. Sund and H. Theorell, in "The Enzymes," P. D. Boyer, H. Lardy, and K. Myrbäck, Eds., Vol. 7, 2nd ed., Academic Press, New York, 1963, p. 25.
82. D. E. Drum, T. K. Li, and B. L. Vallee, *Biochemistry*, 8, 3792 (1969).
83. D. Lindman, M. Zeppezauer, and A. Akeson, *Biochim. Biophys. Acta*, 257, 173 (1972).
84. H. Theorell and B. Chance, *Acta Chem. Scand.*, 5, 1127 (1951).
85. C. C. Wratten and W. W. Cleland, *Biochemistry*, 2, 935 (1963).
86. S. A. Bernhard, M. F. Dunn, P. L. Luisi, and P. Schack, *Biochemistry*, 9, 185 (1970).
87. M. F. Dunn and S. A. Bernhard, *Biochemistry*, 10, 4569 (1971).
88. P. L. Luisi and R. Favilla, *Biochemistry*, 11, 2303 (1972).
89. J. T. McFarland and S. A. Bernhard, *Biochemistry*, 11, 1486, (1972).
90. L. J. Banaszak, D. Tsernoglou, and M. Wade, in *The Structure and Function of Macromolecules and Membranes*, Academic Press, New York, 1971.
91. R. Gerding and R. Wolfe, *J. Biol. Chem.*, 244, 1164 (1969).
92. L. Allen, J. Vanecek, and R. Wolfe, *Arch. Biochem. Biophys.*, 143, 166 (1971).
93. T. Devenyi, S. Rogers, and R. Wolfe, *Nature*, 210, 489 (1966).
94. C. Thorne and N. Kaplan, *J. Biol. Chem.*, 238, 1861 (1963).
95. D. Raval and R. Wolfe, *Biochemistry*, 1, 1118 (1962).
96. B. E. Glatthaar, L. J. Banaszak, and R. A. Bradshaw, *Biochem. Biophys. Res. Commun.*, 46, 757 (1972).
97. K. Harada and R. G. Wolfe, *J. Biol. Chem.*, 243, 4123 (1968).
98. K. Harada and R. G. Wolfe, *J. Biol. Chem.*, 243, 4131 (1968).
99. S. F. Velick and C. Furfine, in "The Enzymes," P. D. Boyer, H. Lardy, and K. Myrback, Eds., Vol. 7, Academic Press, New York, 1963, p. 243.
100. W. Boers, C. Oosthuizen, and E. C. Slater, *Biochim. Biophys. Acta*, 250, 35 (1971).
101. A. Conway, and D. E. Koshland, Jr., *Biochemistry*, 7, 4011 (1968).
102. J. J. M. De Vijlder and E. C. Slater, *Biochim. Biophys. Acta*, 167, 23 (1968).
103. D. Givol, *FEBS Letters*, 5, 153 (1969).
104. R. A. McQuarrie and S. A. Bernhard, in *Pyridine Nucleotide Dependent Dehydrogenases*, H. Sund Ed., Springer Verlag, Berlin, 1970, p. 187.
105. R. A. McQuarrie and S. A. Bernhard, *J. Mol. Biol.*, 55, 181 (1971).
106. J. I. Harris, B. P. Meriwether, and H. H. Park, *Nature*, 198, 154 (1963).

107. D. R. Trentham, *Biochem. J.*, **109**, 603 (1968).

108. T. Keleti in *Pyridine Nucleotide Dependent Dehydrogenases*, H. Sund Ed., Springer Verlag, Berlin, 1970, p. 104.

109. K. Kirshner, *J. Mol. Biol.*, **58**, 51 (1971).

110. N. B. Novoa and G. W. Schwert, *J. Biol. Chem.*, **236**, 2150 (1961).

111. H. Gutfreund, R. Cantwell, C. H. McMurray, R. S. Criddle, and G. Hathaway, *Biochem. J.*, **106**, 683 (1968).

112. G. W. Schwert, and A. D. Winer, in *The Enzymes*, P. D. Boyer, H. Lardy, and K. Myrback, Eds., Vol. 7, Academic Press, New York, 1963, p. 127.

113. M. J. Adams, M. Buehner, K. Chandrasekhar, G. C. Ford, M. L. Hackert, A. Liljas, P. Lentz, Jr., S. T. Rao, M. G. Rossmann, I. E. Smiley, and J. L. White, in *Protein-Protein Interactions*, R. Jaenicke and E. Helmreich, Eds., Springer Verlag, Berlin, 1972, p. 183.

114. M. G. Rossmann, M. J. Adams, M. Buehner, G. C. Ford, M. L. Hackert, P. J. Lentz, Jr., A. McPherson, Jr., R. W. Schevitz, and I. E. Smiley, *Cold Spring Harbor Symp. Quant. Biol.*, **36**, 179 (1972).

115. E. L. Smith, M. Landon, D. Piszkiewiez, W. C. Brattin, T. J. Langeley, and M. D. Melamed, *Proc. Nat. Acad. Sci. U.S.*, **67**, 724 (1970).

116. A. diFranco and M. Iwatsubo, *Eur. J. Biochem.*, **30**, 517 (1972).

117. A. diFranco, *Mecanisme d'action de la glutamate deshydrogenase*, Doctoral Dissertation, University of Paris-Sud, 1971.

118. J. A. Shafer, E. Chiancone, K. L. Yielding, and E. Antonini, *Eur. J. Biochem.*, **28**, 528 (1972).

119. B. L. Horecker, T. Cheng, and S. Pontremoli, *J. Biol. Chem.*, **238**, 3428 (1963).

120. K. Brand, O. Isolas, and B. L. Horecker, *Arch. Biochem. Biophys.*, **130**, 521 (1969).

121. S. Blanquet, G. Fayat, J. P. Waller, and M. Iwatsubo, *Eur. J. Biochem.*, **24**, 461 (1972).

122. A. Levitzki, W. B. Stallcup, and D. E. Koshland, Jr., *Biochemistry*, **10**, 3371 (1971).

123. S. S. Tate, F. Y. Leu, and A. Meister, *J. Biol. Chem.*, **247**, 5312 (1972).

124. M. H. O'Leary and F. H. Westheimer, *Biochemistry*, **7**, 913 (1968).

125. W. Tagaki, J. P. Guthrie, and F. H. Westheimer, *Biochemistry*, **7**, 905 (1968).

126. M. Lazdunski and M. Delaage, in preparation.

127. J. D. Shore and H. Gutfreund, *Biochemistry*, **9**, 4655 (1970).

128. J. J. Holbrook, and R. G. Wolfe, *Biochemistry*, **11**, 2499 (1972).

129. J. J. M. De Vijlder, W. Boers, A. G. Hilvers, B. J. M. Harmsen, and E. C. Slater, in *Pyridine Nucleotide-Dependent Dehydrogenases*, H. Sund, Ed., Springer Verlag, Berlin, 1970, p. 233.

130. W. W. Cleland, *Ann. Rev. Biochem.*, **36**, 77 (1968).

131. W. W. Cleland, in *The Enzymes*, P. D. Boyer, Ed., Vol. 2, 3rd ed., Academic Press, New York, (1970), p. 1.

132. K. Dalziel, *FEBS Letters*, **1**, 316 (1968).

133. H. Buc, personal communication.
134. J. P. Changeux and M. M. Rubin, *Biochemistry*, 7, 553 (1968).
135. A. Levitzki and D. E. Koshland, Jr., *Biochemistry*, 10, 3365 (1971).
136. D. E. Koshland, Jr., *J. Cell. Comp. Physiol.*, 54, 235 (1959).
137. G. N. Cohen, *Current Topics Cell. Regulation*, 1, 183 (1969).
138. L. J. Reed and D. J. Cox in *The Enzymes*, P. D. Boyer, Ed., Vol. 1, 3rd ed., Academic Press, New York, p. 313.
139. A. Ginsburg, and E. R. Stadtman, *Ann. Rev. Biochem.*, 34, 429 (1970).
140. F. Lynen, *New Perspectives Biol.*, 4, 132 (1970).
141. C. Yanofsky and I. P. Crawford, in *The Enzymes*, P. D. Boyer, Ed., Vol. 7, 3rd ed., Academic Press, New York, (1972), p. 1. ˙

RECENT ADVANCES IN
THE CHEMICAL MODIFICATION
AND COVALENT STRUCTURAL
ANALYSIS OF PROTEINS

ROBERT L. HEINRIKSON AND KARL J. KRAMER

Department of Biochemistry
University of Chicago
Chicago, Illinois

141

1 INTRODUCTION

One of the most fascinating aspects of proteins in relation to the other macromolecular components of living systems is their wide-ranging structural and functional diversity. From relatively simple configurations of the structural proteins such as keratin, silk, and collagen, to the highly convoluted folding patterns of enzymes, hormones, antibodies, and other recognitional polypeptides, the proteins comprise a vast spectrum of conformational complexity and biological activity. Recent years have witnessed the growth of a new field of inquiry, in which attempts are made to correlate precisely the unique physical, chemical, and functional properties of a given protein with particular aspects of its structure.

One very productive avenue of approach to elucidating structure-function relationships has been to note the effects on biological activity of specific covalent alterations in the native structure of the protein in question. Interpretation of results thus obtained rests in large measure upon knowledge of the complete covalent structure, but until somewhat more than a decade ago, such information was almost nonexistent. Early reviews on the subject of protein modification [1, 2] indicate a considerable degree of sophistication with regard to reagents then available for selectively altering amino acid side chains. Nevertheless, prior to the widespread application of rapid and accurate methods for amino acid analysis, the significance of such modifications was difficult to assess. A noteworthy example may be found in early chemical modification studies of bovine pancreatic ribonuclease, in which inactivation of the enzyme by iodoacetate, a reagent known to be "specific" for sulfhydryl groups, was assumed to result from the alkylation of an essential cysteine residue [3]. It was not discovered until several years later, when the amino acid sequence was known [4-8] and automated ion-exchange chromotographic amino acid analysis [9-11] had become a well-established procedure, that the loss of activity resulted from carboxymethylation of either His-12 or His-119 in the polypeptide chain [12-14]. Accordingly, the rapid expansion in our knowledge of protein sequences [15], which began in the early 1960s and which resulted directly from the development and application to protein chemistry of high-resolution analytical methods, generated a corresponding growth in the area of chemical modification studies. It became increasingly apparent that careful consideration must be given to questions regarding side reactions and specificity of reagents commonly employed for the selective modification of amino acid residues. Moreover, evidence for the direct functional participation of a certain side chain based upon correlations between its covalent alteration and loss of biological activity was more critically evaluated in terms of possible indirect conformational changes produced by the modification. Eventually, attention began to be focused upon reagents with structural properties similar to the natural ligand, and an increasing number of these so-called affinity labels have been employed for the modification of residues in, or proximal to, the functional sites of enzymes, antibodies, and other recognitional proteins. Reagents with ionization, dipole, or spectral properties sensitive to pH or polarity of the medium were also developed as probes to monitor the local environment of restricted areas of the protein tertiary structure. Bifunctional compounds were investigated as a means of establishing proximity relationships between segments of the polypeptide backbone by cross-linking reactive side chains. These studies directed toward tertiary structure and conformational relationships were complemented by a search for specific modifications which would be useful in elucidating the amino acid sequence. In particular, modifications designed to alter the course of proteolytic cleavage and to facilitate or actually bring about the specific chemical cleavage of the protein have been most successful. In view of the recent spectacular achievements of x-ray crystallography in providing the three-dimensional crystal structures of several proteins, the element of design in the choice of reagents for a particular modification has become even more accentuated. It is indeed a most gratifying chapter in the development of our understanding of the proteins that predictions based upon the chemical approach relative both to primary and tertiary structure concerning groups essential for function are in close agreement with the results obtained by the x-ray crystallographic method. With covalent and crystal structures as a basis for interpreting the results of chemical modifications, and a battery of new physical methods for monitoring signals from native protein or introduced chromophores during catalysis and conformational transitions, the search for correlations between the structure of a protein and its biological function has become a truly meaningful undertaking.

Clearly, protein chemistry encompasses several well-established and expanding disciplines directed toward the eventual understanding in molecular terms of how proteins function. An overall survey of the latest important developments in each of these fields would be inappropriate for the purposes of the present volume, and in fact unnecessary in view of the many excellent texts and review articles related to these individual subjects that have appeared in the recent literature. These may be classified, by and large, in terms of emphasis on modifications related to structural analysis, approaches to the identification of groups in the protein that are important to function, or general methods for selective side-chain modification. Several recent texts have been devoted to the subject of protein chemistry and covalent structural analysis [16, 17]. In particular, two volumes of *Methods in Enzymology*, edited by Hirs [18] and by Hirs and Timasheff [19], have been of invaluable aid to workers in the field. In addition to a comprehensive description of the many methods currently employed in sequence analysis, these sources provide detailed experimental procedures for numerous related chemical modifications. Part of the excellent article by Stark [20] describes some of the more productive approaches to the covalent analysis of proteins, including automated methods which have been applied recently to studies of this kind. In the same source, Spande et al. [21] present a most thorough account of methods available for the specific chemical cleavage of polypeptides. Although the attempts of protein chemists to define elements of tertiary conformation by means of covalent alterations in structure are fragmentary relative to the definitive results of x-ray crystallographic analysis, the chemical approach has been, and will continue to be, useful for the large number of proteins for which no x-ray models are available. References for chemical modification studies related to the conformational analysis of proteins are included in reviews by Habeeb [22] and Brand and Gohlke [23]. General approaches to selective side-chain modification with incisive evaluations of side reactions and pitfalls that may be encountered in the interpretation of cause-and-effect relationships may be found in the articles by Cohen [24] and Glazer [25] and the recent text by Means and Feeney [26]. The Stark review [20] concentrates on those modifications that have been most useful and selective for a given amino acid. Some overlap of reference material along these lines occurs in the articles by Vallee and Riordan [27] and Riordan and Sokolovsky [28], which are largely concerned with modifications of side chains in the protein that are essential for catalysis. Functional-group indentification by unique group, differential, and affinity labeling techniques, and an evaluation of the parameters useful in defining functional sites, are discussed in an earlier review by Singer [29], and Baker [30] has devoted a text to this subject, including methods for the synthesis of active-site-directed compounds. The field of peptide synthesis, which has expanded so rapidly with the application of the Merrifield solid-phase synthetic method (see [31] for a review and references), has contributed in new ways to our understanding of structure-function relationships by examining the regeneration of activity in defective proteins when required synthetic segments of the polypeptide chain are added to the reaction system. This approach is still in its infancy, but it has been applied with success to ribonuclease [32] and staphylococcal nuclease (see [33]). A cursory examination of the table of contents of the latest issues of *Annual Review of Biochemistry* will reveal a host of reviews devoted to structure-function correlations in a variety of proteins, as well as in nucleic acid–protein complexes. Articles by Dickerson [34] and by Hess and Rupley [35] employ as a framework the results of x-ray crystallographic data reinforced by information from chemical methods, and reviews by Scanu and Wisdom [36] and by Marshall [37] discuss the current status in the fields of lipoproteins and glycoproteins, respectively.

From the foregoing, it is clear that there exists a wealth of current reference material dedicated to unifying information from physical, chemical, and biological studies of proteins into some kind of enlightened statement with regard to how all this might explain the molecular mechanisms by which proteins function. For a handful of proteins at least, we are nearing that goal with startling success, and it is entirely reasonable to expect that the future will bring continued advances. Although the contributions of x-ray crystallography cannot be overestimated and this approach will hopefully play an increasingly important role in furthering our understanding of the proteins, the job of the chemist is by no means finished. Interpretation of crystallographic data still relies heavily on knowledge of the covalent structure, and sequence analysis continues to be an important aspect of protein chemistry. Furthermore, although considerable success has been achieved in the crystallographic analysis of proteins for which no sequence data are available, it is very difficult, if not impossible, to identify catalytically, or otherwise, functionally essential groups in the x-ray models in the absence of chemical evidence implicating them as such. Protein chemistry will also undoubtedly expand along lines related to the design of reagents for heavy-atom isomorphous replacement at specific sites and also for synthesizing reagents for cross-linking particular proteins based upon predictions of the x-ray models. Therefore covalent structural analysis and side-chain modifications will continue to be a major preoccupation of the protein chemist, and these topics comprise the subject of this chapter.

2   RECENT ADVANCES IN THE CHEMICAL MODIFICATION OF AMINO ACID SIDE CHAINS

The past few years have witnessed the development of a large number of new reagents for side-chain modifications of proteins. Few, if any, of these new reagents were designed to create an entirely novel experimental approach to the basic questions of structure-function relationship in proteins. Rather, they provide the protein chemist with new tools to refine the established experimental techniques in terms of better quantitation and improved specificity, and to extend this methodology to new and unexplored recognitional and structural proteins.

Traditionally, chemical modifications of side chains in proteins have been carried out (1) to induce changes in, or to abolish altogether, biological activity; (2) to change the conformation or the physical properties; or (3) to introduce special-purpose groups into the protein. Depending on the reagent, these protein modifications can serve various specific aims, such as:

a. Labeling a protein with radioactivity to aid in the location of a particular residue or protein.

b. Reversibly modifying an active site residue, thereby causing inactivation and stabilizing the molecule during purification. Such an approach is often used with proteolytic enzymes that are susceptible to autolysis. The reversible modification allows one to regenerate the native protein at the proper time. This approach is also useful in assessing the functional importance of side chains with reactivity lower than that of a "hyperactive" component, by blocking the latter reversibly.

c. Introducing a "reporter group" to probe the microenvironment of the modified site.

d. Decreasing or increasing the bulk, hence the steric hindrance, of a particular site.

e. Converting one amino acid residue into another. For example, a lysinelike side chain can be formed from cysteine by treatment with ethylenimine. In this way, new sites can be created that are susceptible to tryptic hydrolysis.

f. Cross-linking groups with multifunctional reagents to stabilize tertiary or quaternary structure, to determine intramolecular distances between residues, or to attach the protein to a water-insoluble support.

g. Modifying a group to change its polarity, to stabilize the molecule, or to increase solubility.

As to the stoichiometry of the side-chain substitution, several types of modified proteins can be prepared, each with a particular purpose in mind. Many of the approaches outlined below have been treated extensively in the review by Singer [29].

1. General labeling: The reagent is specific for a certain type of amino acid residue. One chooses the reaction conditions and intrinsic reactivity of the reagent in such a way as to modify all available side chains bearing the same functional group.

2. Affinity or positive labeling: The reagent is designed to possess substrate- or ligandlike features which localize the modification at or near the active center.

3. Differential or negative labeling: A general labeling is first done in the presence of substrate, inhibitor, or ligand, and, after removal of the same, the procedure is repeated with radiolabeled reagent or a different reagent with similar side-chain specificity.

4. Trace or competitive labeling: This approach in effect increases the specificity of the reagent by limiting its reaction to the most labile functional groups. In the presence of a trace amount of radioactive label, the various reactive groups of a protein compete for

the label, and the amount incorporated into any one group is determined by the relative reactivity of that group under the conditions of the experiment. Furthermore, this technique allows easy comparison of the reactivities with that of a model compound added as an internal standard.

The choice of a side-chain-modifying compound is conditioned by several criteria which relate primarily to the application of organic reactions to biological systems. Optimally, the reagent must react in aqueous medium. Many reagents are not readily soluble in water, and to increase their concentration, a small portion of an inert organic solvent can be added, such as dimethyl sulfoxide, dioxane, or acetonitrile. In some cases acetone or dimethylformamide or even methanol, ethanol, ethylene glycol, or acetic acid can be employed for this purpose. When the effects of the modification are to be interpreted relative to the function of the protein, the reagent must be reactive at a pH and at a temperature where the protein is stable and biologically active. Last, but not least, an ideal modifying reagent should be specific, that is, it should react stoichiometrically with the target function with the exclusion of all other functional groups and yield a single, chemically well-defined product. This ideal condition is rarely fulfilled by any reagent, since incorporation of several identical functional groups into the protein matrix usually results in a spreading of reactivities. Some amino acid residues in proteins possess unexpectedly high reactivities, perhaps as a result of their presence at the active site. Conversely, because of the size and the tertiary structure of protein molecules, many of the residues are shielded from solvation ("buried") and relatively unreactive toward reagents dissolved therein. A comprehensive discussion of these factors may be found in the review by Cohen [24]. From data obtained by x-ray crystallographic analysis of 11 proteins, the average exposure of side-chain atoms to solvent was calculated, which reflects the degree of "buriedness" for each type of amino acid residue (Table 1). As expected, hydrophilic residues are solvated more extensively, and hydrophobic side chains are generally much less accessible to solvent. The large disparity of reactivities of identical functional groups in the same protein can be advantageous when probing the microenvironment of individual amino acid side chains. By determination of the relative reactivities of various residues toward several reagents, we can hope to gain some idea as to the positions of the residues in the three-dimensional structure of the molecule, and to understand better the chemical basis for biological function.

**TABLE 1   EXPOSURE AND AVERAGE PERCENT COMPOSITION
OF AMINO ACIDS IN PROTEINS**

Residue	Fractional Exposure[a]	Percent Composition[b]
Alanine	0.32	8.1
Serine	0.43	7.8
Glycine	0.33[c]	7.6
Leucine	0.10	7.3
Lysine	0.58	7.0
Valine	0.16	6.9
Threonine	0.37	6.5
Proline	0.40	5.5
Glutamic acid	0.49	5.3
Aspartic acid	0.42	5.2
Asparagine	0.43	4.8
Isoleucine	0.14	4.6
Glutamine	0.43	4.1
Arginine	0.43	3.9
Phenylalanine	0.17	3.5
Tyrosine	0.25	3.4
Cysteine	0.13	3.4
Histidine	0.27	2.2
Methionine	0.11	1.6
Tryptophan	0.11	1.2

[a] From Rupley and Shrake [38]. X-ray crystallographic data averaged over
11 proteins for all residues. Fractional exposure is the exposure in folded
molecule divided by exposure in unfolded model.
[b] From Dayhoff and Hunt [39]. Data from a pool of 108 protein sequences.
[c] Exposure of atoms in the polypeptide backbone.

Although most protein reagents react with more than one
side-chain functional group, the unique chemical properties of the
many functional groups provide a basis for selective modification. By
taking advantage of the differences in modification rates of various
groups under certain experimental conditions, it is often possible to
bring about a selective modification more or less as desired.
Experimental parameters that may be adjusted to optimize such a
condition include pH, reagent structure and concentration, ionic
strength, temperature, and solvent.

The functional groups that are susceptible to selective chemical
modification in proteins are limited in number. They comprise the α-
and ϵ-amino groups, α-, β-, and γ-carboxyl groups, the guanidino

function of arginine, the hydroxyl groups of serine and threonine, the imidazole ring of histidine, the indole ring of tryptophan, the phenol ring of tyrosine, the sulfhydryl and disulfide groups of cysteine and cystine, and the thioether group of methionine. The frequency at which these residues occur, calculated from the amino acid compositions of 108 proteins, is given in Table 1. The probability of modifying a single group in any class is to a first approximation inversely proportional to the frequency. Basically, most of the functional groups are characterized by the same type of reactivity; they are potential nucleophiles and proton acceptor/ donors. This similarity renders quite difficult the development of group-specific reagents. Table 2 compares the range of acidity determined for several side-chain groups in proteins with the pK values expected in each case from studies of small-molecule analogs. The pK_a values found in proteins give a rough idea about the spread in intrinsic nucleophilic reactivity displayed by these functional groups as a consequence of the three-dimensional conformation of polypeptides. By changing the state of ionization in aqueous medium, one can of course change the apparent reactivity of a functional group by using the proton of the solvent as a blocking agent. Furthermore, steric effects by neighboring groups can affect the reactivity of functional groups [41], as well as that of the reagent itself. For structurally similar amino acids, relative reactivities obey a linear free-energy equation which takes into

TABLE 2   pK VALUES OF TITRATABLE GROUPS IN PROTEINS[a]

Group	Residue	pK_a (Expected)	pK_a (Found)
α-Carboxyl	Carboxyl terminus	2.1–2.4	
β-Carboxyl	Aspartate	3.7–4.0	<1–6.8
γ-Carboxyl	Glutamate	4.2–4.5	
Imidazole	Histidine	6.7–7.1	6.4–7.5
Thiol	Cysteine	8.8–9.1	8–9.5
α-Amino	Amino terminus	7.6–8.0	7.3–>12
ε-Amino	Lysine	10.4	
Phenol	Tyrosine	9.7–10.1	9.4–>12
Guanidino	Arginine	>12	11.5–>12

[a] Condensed from Steinhardt and Beychok [40].

account polar and steric effects [42, 43]. For the reactivity of the same side-chain functional groups in large protein molecules, a quantitative analysis is more difficult. Relative reactivities of groups can be affected positively or negatively by anchimeric effects. Nevertheless, semiquantitative information about relative reactivities of specific residues in proteins can be obtained by isolating peptides from partially modified proteins and comparing relative yields of each (see Sections 2.1 and 2.2). This information in turn helps in the choice of the most appropriate reagent and reaction conditions to achieve the desired selectivity toward a given functional group.

In spite of the bewidering variety of group-specific reagents developed and applied to different proteins, the large majority of these can be classified into relatively few types of reaction mechanisms, since the functional groups on the side chains of a protein are limited in number and the reactivity they display is in no way different from what is observed in organic reactions. Thus an amino, hydroxyl, or thiol group in a protein always obliges by reacting as an electron donor, and the carbonyl function acts as an electron acceptor. In addition to these heterolytic reactions, several oxidative and reductive reagents have also been used in protein modifications, the reactions of which are beyond the scope of this discussion. The following subsections describe the heterolytic-type modifications of each functional group in amino acid side chains. As mentioned in Section 1, the field of protein modification has been the subject of extensive review in recent years [18–30], and the following treatment is limited accordingly to the newest or most important modifications for each particular functional group, subject to the qualification mentioned above.

2.1 Amino Groups

Lysine is one of the major constituents of proteins and is usually located on the "surface" of the molecule, in contact with solvent and readily accessible to chemical agents (Table 1). In addition to the ϵ-amino of lysine, the only other primary amine function occurring in proteins is the α-amino group of the NH_2-terminal residue. The basic forms of these amines are potent nucleophiles through the intermediacy of the unshared pair of electrons of the nitrogen atom. The ϵ-amino group is a strong base of $pK_a = 10.2$, whereas the α-amino group is more acidic, as shown by its $pK_a = 7.8$ (Table 2). Because of these ionization characteristics, the general-purpose modifications of amino groups are usually carried out at high pH to ensure at least partial deprotonation of the ϵ-amino groups. Selectivity toward the α-amino group can be approached by working at lower pH values, such as pH 7, and by using modifying reagents of relatively low reactivity.

Amino groups readily undergo acylation, alkylation, arylation, and deamination reactions (Table 3). The chemical modification of amino groups has been used most productively in covalent structural analysis, both in sequential degradation from the NH_2-terminus (Edman degradation) and in the blocking of lysine side chains to obtain overlaps of tryptic peptides.

A typical reagent that displays high selectivity for amino groups is 2,4,6-trinitrobenzene sulfonate (TNBS) [44]. Arylation by this reagent is carried out at pH 9.5. Sulfite is displaced from TNBS by the amine nucleophile to give the trinitrophenylated secondary amine derivative. The reaction may be monitored and quantitated spectrophotometrically, hence TNBS is a particularly convenient analytical probe for assessing the total amino group content of the protein.

Acetylations of amino groups are rapid, mild, and relatively group-specific [45]. Acetic anhydride was used by Hartley and co-workers to determine the relative reactivities of the amino groups of elastase by a trace labeling technique [46, 47]. Their method consists of treating a protein with a limiting amount of radioactive reagent and then with an excess of unlabeled reagent to yield a chemically homogeneous but heterogeneously labeled protein. After suitable proteolysis, peptides containing each labeled group were isolated to determine specific radioactivities, and therefore relative reactivities. Rate constants for acetylation were calculated relative to that of the amino group of phenylalanine, and differences in rates were attributed to steric and polar factors. In elastase, Lys-87 and Lys-224 were both found to have a pK_a = 10.3 and normal reactivity, while the amino terminus had a pK_a = 9.7 and abnormally low reactivity. More recently, 3-acetoxy-1-acetyl-5-methylpyrazole has been used to introduce the acetyl group into proteins [48].

Anhydrides of dicarboxylic acids are also amino group–specific reagents; they produce a charge reversal at the modified residue, since the positively charged ammonium ion is replaced by a negatively charged carboxylate [49–51]. Shifting the isoelectric point by charge reversal often increases the solubility of sparingly soluble proteins and peptides, and it can even cause dissociation of oligomeric proteins into subunits. Another great advantage of the use of anhydrides of dicarboxylic acids resides in the reversibility of the modification under moderately acidic conditions by nucleophilic attack of the free carboxyl function on the amide group of the substituent. In addition to the often used succinic anhydride, the related compounds maleic and methylmaleic (citraconic) anhydride have found recent application in structural studies. Removal of the substituent under acidic conditions is much more rapid with the last-mentioned two compounds.

Acylation with N-acetylhomocysteine thiolactone results in the introduction of a thiol function at the locations of the α- and ε-amino groups [52]. Proteins so modified may provide interesting derivatives for studying aggregation reactions, scrambling of disulfide bonds, and perhaps the creation of new catalytic sites for hydrolytic enzymes.

Guanylation of the lysine residues preserves the positive charge of the side chain by producing homoarginine [53]. O-Methylisourea and 1-guanyl-3,5-dimethylpyrazole (GDMP) have been used for this purpose at pH 10.5 and pH 9.5, respectively. Presumably because of this difference in pH, GDMP appeared to be more selective toward the ε-amino group. The reaction of amino groups with imido esters such as methyl acetimidate and dimethyl suberimidate to form amidines also preserves the positive charge in lysine [54]. The advantage of the use of these reagents resides in the fact that the modifications may be performed at a lower pH (7–10), and that the reagents themselves are more specific for amino groups than are the guanylating agents. Diimido esters can be used for inter- and intramolecular cross-linking of polypeptides via lysyl residues [55].

Pyrylium salts have been employed as specific reagents for lysine residues [56]. These compounds show an apparent steric selectivity, since they form the corresponding N-substituted pyridinium salts with only a few of the total number of available side chains (Scheme 1).

Further progress in the chemical modification of the amino group can be foreseen in several directions. The presently available

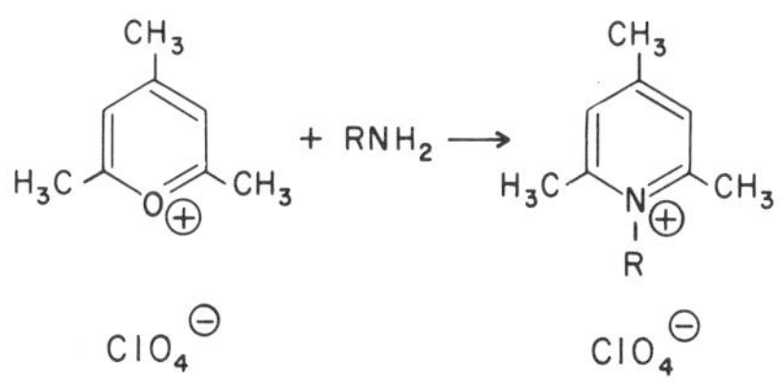

Scheme 1

TABLE 3   CHEMICAL MODIFICATION OF AMINO GROUPS

Reagent	Protein	Results	Reference
2,4,6-Trinitrobenzenesulfonic acid	Tobacco mosaic virus coat protein	Amino groups quantitated	57
	Elastase	Lysines and tyrosines reacted	58
	Serum albumin and ribonuclease	Lysines trinitrophenylated	59
	Pyruvate kinase	Four lysines modified; inactivation	60, 61
	Insulin, ribonuclease A and T_1	Lysines quantitated	62
	Glutamate dehydrogenase	Lys-428 and Lys-425 reacted; inactivation	63–65
Acetic anhydride	Elastase	Lys-87 and Lys-224 acetylated normally; NH_2 terminus acetylated slowly	46
	Erythrocyte membrane proteins	Calcium binding increased	66
	Chymotrypsin	NH_2 terminus acetylated	47
Maleic anhydride	Elastase	Inactivation	67
Succinic anhydride	Chymotrypsin	Decreased activity	68
	Transferrin	60 of 70 amino groups acetylated in native	69
	Inorganic pyrophosphatase	Subunit dissociation; limitation of proteolysis	70
	Chymotrypsinogen	4 of 6 lysines succinylated	71
Citraconic anhydride	Immunoglobulin G, lysozyme, pepsinogen	All lysines acetylated; regenerated at acidic pH	72
			73
			74
N-Acetylimidazole	Thermolysin	Lysines and tyrosines reacted; partial inactivation	75

154

Reagent	Protein	Description	Ref.
3-Acetoxy-1-acetyl-5-methyl-pyrazole	Ribonuclease A, bovine serum albumin, ovalbumin, chymotrypsin, lysozyme, trypsin, insulin	Amino groups acetylated; hydroxyl groups also reacted	48
N-Acetylhomocysteine thiolactone	γ-Globulin	Converted lysines to N-acetylhomocysteinyl derivatives	76
O-Methylisourea	Bovine trypsin inhibitor	Lys-18, Lys-66, N-terminal Phe modified; inactivation	77
	Staphylococcal enterotoxin	All lysines modified; active	78
	Deoxyribonuclease	Lysines guanidinated; full activity	79
	Naja neurotoxin	Lysines modified; active	80
	Soybean trypsin inhibitor	Lysines guanidinated; full activity	81
	Ribonuclease A	All 10 lysines modified; inactivation	82
1-Guanyl-3,5-dimethyl-pyrazole	Trypsin	10 of 15 lysines modified; active	83
	Trypsinogen	All 16 lysines modified; could be fully activated	84
Methyl acetimidate	Prophospholipase A	Lysines amidinated; could be activated	85
Methyl picolinimidate	Liver alcohol dehydrogenase	50 of 60 lysines modified; V_{max} increased	86
2,4,6-Trimethylpyrylium perchlorate	Chymotrypsin	6 lysines modified	56
	Acetoacetate decarboxylase	1 lysine modified; inactivation	56
Cyanate	Glutamate dehydrogenase	Lys-97, Lys-85, and α-amino carbamylated, inactivation	87
Nitrous acid	Thrombin	Surface residues reacted; esterase activity decreased	88–90
	Chymotrypsin	α-Amino groups deaminated	91, 92

charge-reversal reagents produce a rather large extension and a shift of the electric charge by several angstroms. A need exists for reagents that would reverse the charge but preserve its approximate location. Also, it would be desirable to introduce ionizable groups with pK values extended over a wide range. As a parallel to normal acylating agents, which always increase the bulk of the side chain at the same time as they neutralize the charge of the amino group, it would be interesting to study the effects of degradation of the lysyl side chain through elimination of the amino function. Furthermore, the involvement of amino groups in the function of several enzymes will certainly stimulate development of active-site-directed reagents for the amino group.

2.2 Carboxyl Groups

Proteins contain three types of carboxyl groups: the α-carboxyl of the carboxyl-terminal residue, the β-carboxyl of aspartic acid, and the γ-carboxyl of glutamic acid. These groups are characterized by intrinsic pK_a values of 2.2, 3.8, and 4.3, respectively, whereas their apparent p$\bar{K}_a$ values in the protein molecule can range anywhere from 1 to 7, depending on whether they are perturbed by electrostatic charges, hydrogen bonds, or internal salt-bridge formation (Table 2). Carboxylates are among the most abundant functional groups in proteins. Glutamic and aspartic acids constitute on the average about 10% of all the residues, and they are generally located at the surface of the molecule, second only to lysines in the degree of exposure (Table 1).

Because of their high frequency and hydrophilicity, protein carboxyl groups can be modified by a variety of reagents, such as those leading to aminolysis (carbodiimide-amine nucleophile and isoxazolium salts–amine nucleophile), esterification (diazonium, oxonium, and isoxazolium salts), and reduction (alkyl boranes). The results of these and other recent carboxyl group modifications are summarized in Table 5.

Perhaps the most popular modification of carboxyl groups is that employing an amine nucleophile reagent and a water-soluble carbodiimide as a catalyst [93]. Scheme 2 shows the reaction pathway proposed by Khorana to account for the numerous products formed by this reaction [94]. The carbodiimide 1, an active electrophile, reacts with the carboxylate ion 3 at slightly acidic pH to give an O-acylisourea 4. This activated intermediate can either rearrange via an O $\rightarrow$ N acyl shift, forming 5, or react with

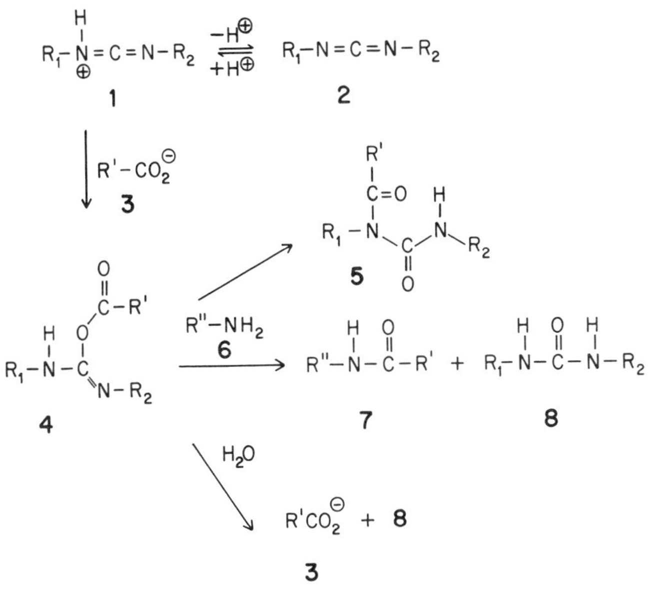

Scheme 2

nucleophiles such as 6 to give the corresponding amide 7. The overall reaction can be conceived of as a hydration of the carbodiimide to the corresponding urea 8 by water released in the amide bond formation.

Hegarty and Bruice [95] studied the mechanism of aminolysis of 2-amino-4,5-benzo-6-oxo-1,3-oxazine (9, Scheme 3), a product of an intramolecular carboxyl addition to a carbodiimide. Formation of the N-acylurea (5, Scheme 2) from compound 9 through an intramolecular O → N shift is sterically prohibited. The cyclic benzoxazine 9 can undergo several ionizations (Scheme 4), yielding

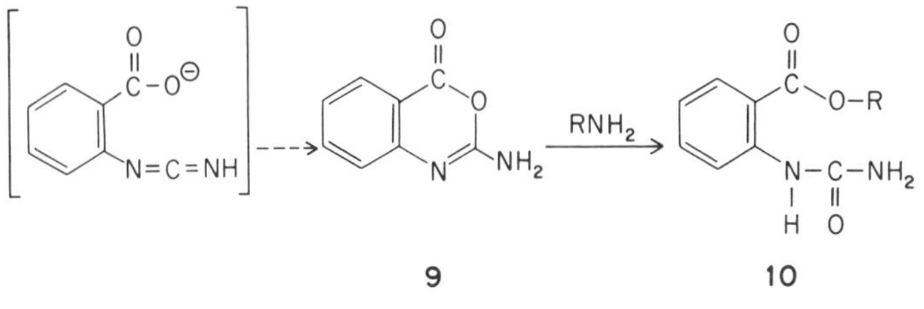

Scheme 3

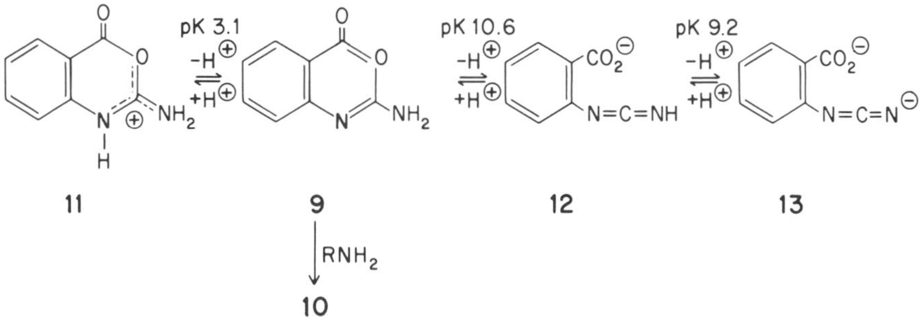

Scheme 4

species that are neutral, monoanionic (12), and dianionic (13). The neutral form **9** is predominant between pH 4 and 10. Although reactions of **9** with hydroxide ion and of **11** with water are significant in basic and acidic solutions, respectively, in the neutral pH region (pH 4 to 8) hydrolysis is slow relative to the formation of the amide **10**. Extrapolating these results to the carbodiimide-catalyzed aminolysis of protein carboxylates, it is evident why the incorporation of the added nucleophile into protein carboxyl groups is highly favored under conditions in which added amine is in large excess over any nucleophilic groups in the protein. Because of their high reactivity toward carboxylates, carbodiimides provide a fairly selective means for the general labeling of all three types of carboxyl groups, although some modifications of phenol, hydroxyl, and thiol groups may also occur [96, 97].

The catalyst 1-ethyl-3-dimethylaminopropylcarbodiimide (EDC) has been used for the aminolysis of the carboxyl groups of lysozyme by sulfanilic acid [98]. Trace labeling was carried out to determine the relative reactivities of all 10 carboxyl groups present in the enzyme. Glu-35 and Asp-101 were notably more reactive than the other groups, and Glu-7, Asp-18, and Asp-66 were modified to the least extent (Table 4). The presence of a competitive inhibitor of the hydrolytic activity of the enzyme significantly retarded the reaction of Glu-35 and Asp-101, two residues known from crystallographic studies to be at the active site [99].

Carboxypeptidase A was inactivated as a result of the modification of Glu-270 following reaction with carbodiimide in the absence of any added nucleophile [100]. An undetermined carbonyl-modified product was formed, most likely the *N*-acylurea (**5**, Scheme 2).

N-Alkyl-5-phenylisoxazolium salts (Woodward's reagent, **14**, Scheme 5) are another type of carboxyl-activating agent frequently used in protein chemistry [101]. The reaction of **14** with carboxylates proceeds through the irreversible formation of a ketoketenimine **15**, structurally related to carbodiimides. The ketoketenimine reacts in turn with a carboxyl group to give a relatively stable enol ester **16**. The enol can then be displaced from the acyl group by an appropriate nucleophile, usually an amine, to yield a stable amide **17**. Two side reactions can complicate this simple reaction scheme:

1. Above pH 3, the ketoketenimine rapidly hydrolyzes to form the corresponding amide **18**.
2. The enol ester can undergo a base-catalyzed rearrangement to the imide **19**.

Modification of carboxylates with the Woodward reagent and

TABLE 4   RELATIVE REACTIVITY OF LYSOZYME CARBOXYL
GROUPS TOWARD SULFANILIC ACID AND EFFECT OF
ADDED COMPETITIVE INHIBITOR[a]

	Fraction Reacted	
Carboxyl	A: Trace labeling 0.05 M sulfanilic acid, 0.05 M EDC, pH 5, 2.4 groups reacted	B: Conditions same as A, except 0.01 M (GlcNAc)₃, 1.2 groups reacted
Glu-7	Low	0.09
Asp-18	0.03	0.02
Glu-35	0.75	0.14
Asp-48, Asp-52[b]	{ Low { 0.28	{ 0.04 { 0.11
Asp-66	Low	Low
Asp-87	0.08	0.07
Asp-101	0.86	0.25
Asp-119	0.23	0.27
α-Carboxyl	0.18	0.22

[a] From Kramer and Rupley [98]. EDC, 1-ethyl-3-dimethylaminopropylcarbodi-imide; (GlcNAc)₃, β-(1→4)-linked trisaccharide of N-acetylglucosamine.
[b] Asp-48 and Asp-52 are present in the same tryptic peptide, and therefore the values shown cannot be assigned to a specific residue.

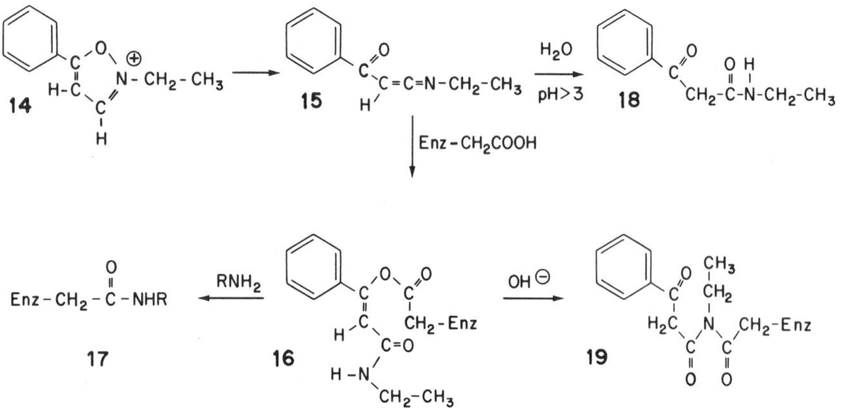

Scheme 5

concommitant loss of enzymic activity provided evidence for the presence of carboxylates at the active sites of trypsin [102], and carboxypeptidase [103, 104].

Active-site-directed aliphatic diazo compounds can also be used for the selective modification of carboxyl groups [105]. The best reagents of this type are the diazoacetyl or the diazomethyl ketone derivatives, because of the stabilizing effect of the carbonyl group on the otherwise extremely unstable diazomethyl function. Even with these stabilized derivatives, selectivity is somewhat less than optimal, and they also readily alkylate sulfhydryl groups. Chemical modification of carboxyl groups with aliphatic diazo compounds has implicated aspartates and glutamates at the active site of several acid proteases (Table 5).

The Meerwein reagent, triethyloxonium fluoroborate, is a highly reactive alkylating agent, and for this reason selective modification of carboxyl groups with this reagent can be achieved with only a few proteins under restricted conditions [105]. In general, methionine, histidine, and cysteine also react readily. In most proteins methionine is buried, and the native conformation usually prevents its reaction with the Meerwein reagent. The side reaction of histidine can be controlled by working under slightly acidic conditions. On the basis of chemical modification of proteins with the Meerwein reagent, carboxyl groups have again been implicated as active-site components of lysozyme [106] and trypsin [107].

An interesting esterification reaction of Glu-35 in lysozyme is that reported by Rupley et al. [99]. In these experiments KI_3 was used

TABLE 5 CHEMICAL MODIFICATION OF CARBOXYL GROUPS

Reagent	Protein	Results	Reference
Carbodiimide-nucleophile	Staphylococcal enterotoxin B	30 of 39 carboxyl groups converted to amides with no loss of activity; inactivity lost after 35 groups modified	110
	Chymotrypsinogen and chymotrypsin	Asp-194 unreactive in both proteins	111–113
	Transferrin	16 of 55 carboxyl groups reacted with no loss of activity	114
	Trypsin	Asp-177 and Asp-182 protected by competitive inhibitor	115
	Insulin	All 6 carboxyl groups reacted in native hormone	116
	Thrombin	All 8 carboxyl groups reacted in native; thorium protected 7 groups	117
	Cobrotoxin	6 of 7 carboxyl groups reacted with no loss of activity; Glu-21 reacted in 5 M guanidine and activity is lost	118
	Albumin	All carboxyl groups reacted in native; S-carboxymethylation decreased extent of reaction	119
	Aspartate aminotransferase	50% of the carboxyl groups reacted in native; activity decreased	120
	Tobacco mosaic virus coat protein	6 carboxyl groups reacted; aggregation property lost	121
	Subtilisin	7 of 16 carboxyl groups reacted in native	122
	Myoglobin	Glu-83 and Glu-85 modified	123
	Pepsin	Inactivation	124

TABLE 5 CHEMICAL MODIFICATION OF CARBOXYL GROUPS (*Continued*)

Reagent	Protein	Results	Reference
Carbodiimide-nucleophile (*continued*)	Collagen	Most carboxyl groups modified; calcifying activity abolished	125
	Glucagon	All 4 carboxyl groups reacted in 7 M guanidine; activity lost	126
	Hemerythrin	8 of 18 carboxyl groups react; dissociation to monomer	127
	Lysozyme	Reactivity of all carboxyl groups determined; Asp-101 and Glu-35 protected by substrate	98
	α-Chymotrypsin	Surface carboxyl groups modified	128
	Trypsin, chymotrypsin, lysozyme, ribonuclease	All carboxyl groups converted to carboxamido groups in 5.5 M NH_4Cl	129
	Adenosine triphosphatase	10% of carboxyl groups reacted in native; inactivation	130
Carbodiimide	Carboxypeptidase A	Glu-270 modified; inactivation	100
	Alkaline proteinase	Some carboxyl groups reacted and elastolytic activity introduced	131
	Adenosine triphosphatase	Inactivation	130
N-Methyl-5-phenylisoxazolium fluoroborate	Trypsin	2 or 3 carboxyl groups converted to enol esters or amides; inactivation; competitive inhibitor protected two groups	101, 102
	Carboxypeptidase A	Several carboxyl groups modified; inactivation; competitive inhibitor protected Glu-270	103, 104
Aliphatic diazo compounds	Pepsin	1 or 2 carboxyl groups converted to ester; inactivation	132

	Cathepsin D	Inactivation	133
	Pepsin C	Inactivation	134
	Pepsin	Ile–Val–Asp–Thr–Gly–Thr–Ser $\overset{|}{\underset{|}{C}}$=O OR	135
	Cathepsin E	Inactivation	136
	Bacterial acid proteases	Inactivation	137
	Rhizopus acid protease	Inactivation	138
	Pepsin	Bifunctional reagent caused cross-linking and inactivation	139
Triethyloxonium fluoroborate	Lysozyme	Asp-101 and Asp-52 esterified; inactivation	106
	Trypsin	2 carboxyl groups ethylated; 80% loss of activity	107
	Pepsin	Two carboxyl groups modified; both essential for activity	140
Alkylboranes	Myoglobin, lysozyme	All or some of the carboxyl groups modified, depending on steric properties of reagent	108
1-Ethoxycarbonyl-2-ethoxy-1,2-dihydroquinoline	α-Chymotrypsin	Single carboxyl modified, probably to anhydride; inactivation	109

to oxidize tryptophan. The oxindole derivative of Trp-108 is formed via an intermediate, the Glu-35 enol ester Trp-108 lysozyme, in which the carboxyl group of Glu-35 is covalently bound to the indole C-2 atom of Trp-108 (Scheme 6). Studies with indoles and

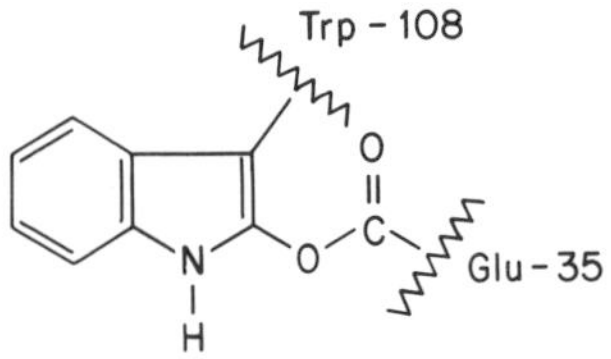

Scheme 6

substituted indoles have shown that the oxidation rates by I$^+$ are strongly enhanced by carboxylates and by general base catalysts, presumably via intermediates analogous to the enol ester derivative of lysozyme.

A class of reductive-type reagents is noteworthy in regard to carboxyl-group modifications. Diborane reduces all the carboxyl groups of myoglobin and lysozyme to the corresponding alcohols at low pH and temperature [108]. Disamylborane and 9-borabicyclono-nane appear to be specific for the reduction of glutamates and the unhindered terminal α-carboxylate. The usually reactive disulfides are not reduced under these conditions, probably because they are buried inside the protein molecule. The remarkable specificity displayed by the alkylboranes for the chain length of the carboxyl-bearing side chain might eventually lead to their use for the determination of the relative amounts of aspartate and glutamate residues in native proteins.

An anhydride forming agent, 1-ethoxycarbonyl-2-ethoxy-1,2-dihydroquinoline, has been used to inactivate α-chymotrypsin at slightly acidic pH, perhaps by forming a stable mixed anhydride at Asp-194 [109].

Because of the large variation in the pK values and in the microenvironments of carboxyl groups in proteins, it is not surprising that selective modification of carboxylates is in a less than satisfactory state. Clear distinction between α-, β-, and γ-carboxyl groups by chemical modification is still a problem of the future. This task is more challenging than the differentiation between the α- and ε-amino groups because of the similarity of the acidities of all carboxylates. Furthermore, in order to modify all the carboxylate groups in a protein, the presence of some unfolding agent is required. Perhaps with the exception of diazomethane, all the reagents currently employed for carboxylate modification are too bulky to overcome the local steric hindrances of the carboxyls in the tertiary structure. Studies directed toward establishing carboxylate groups as functional components of proteins would be facilitated by a new class of carboxyl-specific reagents of small size, which may be more broadly reactive with native proteins.

2.3 Guanidino Groups

The δ-guanidino group of arginine is strongly basic (pK_a = 12.5), and in the protonated form it is resistant to all but the most severe chemical treatments. This functional group is a planar structure stabilized by resonance, and it remains in the protonated form over the usual pH range of protein stability. Arginine is less abundant in proteins than most other amino acids, and it is generally solvated extensively at the surface of the molecule (Table 1). The reagents that show the highest specificity toward the guanidino group of arginine are those containing two adjacent or closely proximal carbonyl groups (Table 6) [141]. Because the distance of the two carbonyl functions closely matches that of the two unsubstituted nitrogen atoms of the guanidino group, 1,2- and 1,3-dicarbonyl compounds readily form heterocyclic condensation products with arginine. These widely employed electrophilic reagents also react to a lesser extent with amines and sulfhydryl groups, yielding Schiff bases and other products, but the proximity effect strongly favors addition to the guanidino function. Thus glyoxal, phenylglyoxal, and 2,3-butanedione react even at neutral pH with an almost exclusive formation of the arginine adduct. Phenylglyoxal has been shown to react rapidly with arginines to produce derivatives that contain two phenylglyoxal moieties per guanidino group (20, Scheme 7) [142]. Under mildly acidic conditions, these products are sufficiently stable to permit isolation of labeled peptides, although arginine can be regenerated at a more alkaline pH in the absence of excess reagent. The product formed in the reaction of 2,3-butanedione with arginine has not been characterized, but the stoichiometry of the reaction corresponds to an overall addition of 3 moles of dione per mole of arginine.

The above-mentioned side reactions with the amino group, which usually accompany the modification of arginine by dicarbonyl reagents at neutral or alkaline pH, may be minimized by carrying out

TABLE 6 CHEMICAL MODIFICATION OF GUANIDINO GROUPS

Reagent	Protein	Results	Reference
Glyoxal	Horse heart metmyoglobin	1.1 arginine residues modified; no inactivation	144
	α-Glucanphosphorylase	Arginine and lysine modified; inactivation; substrate protected	145
Phenylglyoxal	Ribonuclease T$_1$	Arg-77 modified; inactivation	146
	Trypsin	Soybean trypsin inhibitor protected Arg-55	147
	Carboxypeptidase B	One arginine modified; k_{cat} decreased; K_m unchanged	148
	Pancreatic trypsin inhibitor	Arg-1, Arg-17, Arg-20, Arg-39, Arg-42 modified without loss of activity	149
	Lactate dehydrogenase	2 arginines per subunit modified; K_m decreased	150
2,3-Butanedione	Ribonuclease	75% of the arginines modified; 45% of the activity retained	151, 152
	Bovine plasma albumin	93% of the arginines and 30% of the lysines modified; lysines protected with citraconic anyhydride	151, 152
Nitromalondialdehyde	Lactate dehydrogenase	All arginines (3 per subunit) modified	153
	S-carboxymethyl B chain of insulin	All arginines modified	154

166

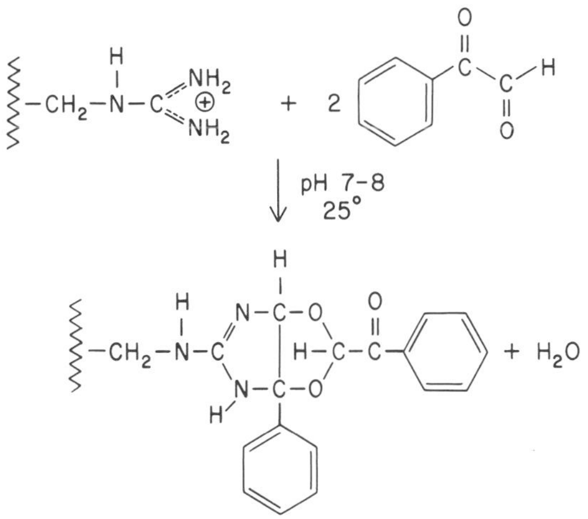

20

Scheme 7

the reaction at pH 6. Van Chuyen et al. [143] studied the reaction of dipeptide amino groups with glyoxal (**21**, Scheme 8) at pH 5 and 100°C. Products detected included carbon dioxide, ammonia, aldehydes, and pyrazinones. These amino group modifications, which occur at higher temperature, may not be important under physiological conditions, but they should be borne in mind as possible side reactions. Side reactions with amino groups can be eliminated altogether, however, by the following multistep modification procedure. First, the amino groups are protected by acylation with citraconic anhydride [51]. Next, the arginines are modified at pH 8 with the dicarbonyl reagent, and finally the amino groups are regenerated by removal of the citraconyl substituents at acidic pH.

The remarkable ability of the guanidino group to form several hydrogen bonds simultaneously with proximally disposed carboxylates seems to play an important role in the stabilization of the tertiary structure of proteins. Development of arginine-specific reagents that would react at one or the other of the two N^{ξ} atoms would certainly help in understanding these chain-chain interactions in proteins and in clarifying the role of arginine in stabilizing quaternary structures.

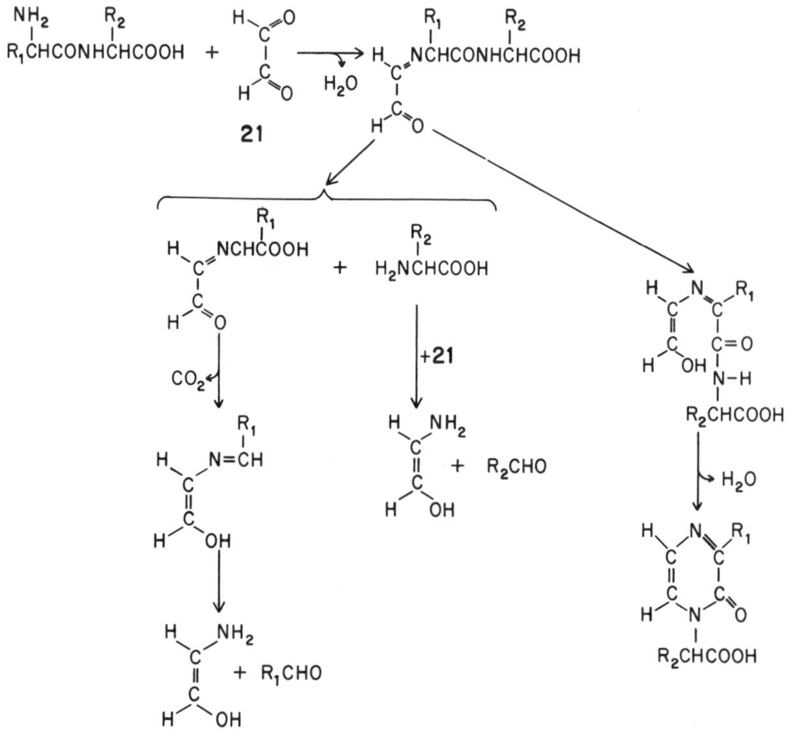

Scheme 8

2.4 Aliphatic Hydroxyl Groups

The aliphatic hydroxyl groups of major importance in proteins are the primary β-hydroxyl of serine and the secondary β-hydroxyl of threonine. In some special proteins, hydroxyproline and hydroxylysine might occur even in relatively large abundance, but this section focuses only upon serine and threonine. The aliphatic hydroxyl groups in proteins are usually well exposed to solvent, and they occur at quite high frequency (Table 1). The hydroxylate ion is strongly basic. It has been estimated that its intrinsic pK in proteins is 12 or higher, and its ionized form has never been convincingly demonstrated in native proteins. Because of this high pK, it is difficult to modify serine and threonine selectively, because many other functional groups with lower pK values show a higher apparent nucleophilicity in the neutral pH range. An important exception to this rule is the highly reactive hydroxyl group at the active site of serine enzymes where a neighboring histidine facilitates acyl transfer reactions at the serine. In spite of the great abundance of reagents that react selectively with the active-site serine [155], no general serine- and/or threonine-specific reagent has been described to date. One might envision a negative labeling approach whereby a strong acylating agent could react with hydroxyl groups after reversible blocking of the more reactive nucleophiles. Since active-site serine-modifying reagents have been extensively reviewed previously, we mention only a few of the recently described reagents (Table 7).

The reactive serine of chymotrypsin was modified with a water-soluble carbodiimide in the absence of added nucleophiles [156]. Chymotrypsin and elastase have also been inactivated by alkyl isocyanates (22, Scheme 9) of various chain lengths [157,

Scheme 9

158]. These reagents are presumably directed toward the active site by binding of their alkyl chain to the hydrophobic cleft of the active site. The enzymes appear to recognize the chain length of the reagent, since chymotrypsin is inactivated most specifically by octyl isocyanate, whereas elastase is inactivated best by butyl isocyanate. Trypsin is not affected by either reagent.

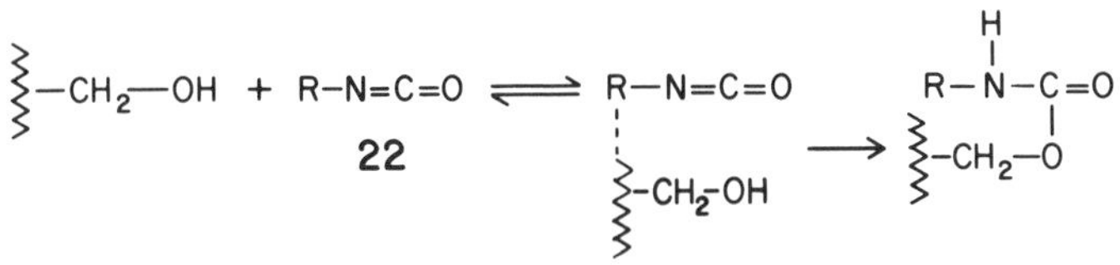

A more general reaction for serine and threonine would be the N → O acyl migration in strongly acidic solution, an unfavorable condition for most proteins [159]. In a study of the esterification of the carboxyl groups of insulin in methanol-HCl, a specific N → O acyl shift was found to occur [160]. The product was an insulin derivative with an ester bond instead of a peptide bond between Tyr-B26 and Thr-B27.

2.5 Imidazole Groups

The imidazole ring of histidine is a relatively large, basic nitrogen heterocycle which, when incorporated into a peptide chain, has an intrinsic pK_a = 6.7. Histidine occurs rather infrequently in proteins

TABLE 7   CHEMICAL MODIFICATION OF ALIPHATIC HYDROXYL GROUPS

Reagent	Protein	Results	Reference
Carbodiimide	α-Chymotrypsin	Active-site serine modified; product unknown	156
Alkylisocyanates	Chymotrypsin; elastase	Active-site serines acylated; inactivation	157, 158
Methanol-HCl	Insulin	N → O acyl migration between Tyr-B25 and Thr-B27	160

(only methionine and tryptophan are less abundant), and the histidyl side chain is on the average 50% shielded from solvent (Table 1). Because of the ionization of the ring at neutral pH, and because of the delocalization of the electric charge by resonance in the ring, the imidazole function is an excellent catalyst of ester hydrolysis. This catalysis occurs not only by the usual pathway of nucleophilic attack of the carbonyl function followed by rapid hydrolysis of the acyl imidazole, but in many cases imidazole acts as a general base catalyst [161]. These unique catalytic properties of histidine are also manifested in enzymic catalysis, and indeed histidine is the catalytic function most frequently found at the active sites of enzymes.

A difficulty in achieving selective modification of histidine stems from the fact that these residues are commonly buried in proteins and participate in intramolecular hydrogen bonding. Although selective modification of histidine by acylation is complicated by the instability of acyl imidazoles, isolation of stable acyl imidazole derivatives of proteins has been achieved by reaction with diethyl pyrocarbonate. Ethoxycarbonyl imidazoles are relatively stable, just as are carbamic esters (urethans) [162]. Diethyl pyrocarbonate has been used to modify histidines in ribonuclease [162], arginine kinase [163], and creatine kinase [163]. Extensive modification of amino groups may also occur with this reagent, as shown in the modification of pepsin [162], trypsin, [162] and phospholipase [164]. However, several recent studies of the modification of histidine in proteins with diethyl pyrocarbonate indicate that, under strictly defined reaction conditions, the reagent has good applicability as a specific reagent for the imidazole group (Table 8). Finally, it should be mentioned that the modification of histidine by this reagent can be reversed by treatment with hydroxylamine as the acyl acceptor.

A site-specific alkylating agent for His-57 of chymotrypsin is methyl-*p*-nitrobenzenesulfonate which inactivates the enzyme by methylation of N-3 of the imidazole ring [165]. Again, under more drastic conditions the selectivity of this reagent is lost, and indeed it can be used as an excellent sulfhydryl-modifying agent [166]. Haloacetates are good general alkylating agents of histidine, although the high reactivity of cysteine, lysine, and methionine with these compounds prevents them from being specific. Ribonuclease is a notable exception to this rule, since it may undergo very specific *N*-alkylation of active-site histidine residues by reaction with haloacetates. At pH 5.5, carboxymethylation of ribonuclease by iodoacetate occurs either at N-1 of His-119, or N-3 of His-12; under

TABLE 8 CHEMICAL MODIFICATION OF IMIDAZOLE GROUPS

Reagent	Protein	Results	Reference
Diethyl pyrocarbonate	Octopine dehydrogenase	2 histidine residues modified; inactivation; reversible with hydroxylamine	171
	Arginine oxygenase	1 histidine modified; inactivation	172
	Liver alcohol dehydrogenase	4 histidine residues modified per subunit; partial inactivation	173
	Ceruloplasmin	Decreased oxidase activity	174
	6-Phosphogluconate dehydrogenase	K_m decreased	175
	Glutamate dehydrogenase	3 of 4 histidines modified; partial inactivation	176
	Crotalus phospholipase	Histidines and lysine ethoxycarbonylated; histidine regenerated with hydroxylamine; inactivation	164
	Glutamate dehydrogenase	1 histidine modified; V_{max} increased; NADH protected	177
	Concanavalin A	2 histidines reacted; metal protected	178
	Glucagon	Amino-terminal histidine reacted; inactivation; hydroxylamine reactivated	179
Methyl-p-nitrobenzenesulfonate	Chymotrypsin	3-Methyl-His-57 produced; inactivation	165
Iodoacetate	α-Lactalbumin	His-32, His-68, His-107, Met-90 modified; decreased activity	180
	Insulin	2 histidines of B chain modified	181
	Deoxyribonuclease II	1 histidine alkylated; inactivation	169

Reagent	Protein	Description	Ref.
Iodoacetamide	Malate dehydrogenase	3 histidines modified; inactivation; no inactivation with iodoacetate	182
Bromoacetate	Myoglobin (crystal)	3-Carboxymethyl-His-36; 1,3-dicarboxymethyl-His-12, -His-81, -His-114, -His-116; small amount of 1-carboxymethyl-His-119; His-24, His-36, His-81, His-93, His-97, and most of His-119 unmodified; a few lysines also alkylated	183
	Myoglobin (solution)	Similar to crystal, except His-36 unreacted and more lysines modified	184
	Carbonic anhydrase B	Histidine modified; inactivation	185
	Ribonuclease A	Reactivity: His-119 (N-1), His-12 (N-3), His-105 and 1-carboxymethyl-His-119 (N-3); methionine and lysine also modified	167
5-Diazonium-1-*H*-tetrazole	Subtilisin	Histidine modified; inactivation	186
	Hemoglobin	28 of 38 histidine residues and 8 of 12 tyrosines modified	187
	Myoglobin	8 of 11 histidines and both tyrosines modified	187
	α-Glycerophosphate dehydrogenase	2 histidines modified; inactivation	188
	RNA polymerase	54 of 60 histidines and 33 of 100 tyrosines modified; inactivation	189
Fluorodinitrobenzene	α-Chymotrypsin	His-57 modified; more reactive than *N*-acetylhistidine by competitive labeling	190

appropriate conditions, no other amino acids are modified [167]. Reactions at these histidines are about 2000 times faster than those of histidine itself. An even more striking example is the inactivation of deoxyribonuclease II by iodoacetate, in which case the reaction is 1000 times faster than that with ribonuclease [169]. In deoxyribonuclease II, a single histidine undergoes carboxymethylation at N-3.

At one time, diazonium-1H-tetrazole was thought to be specific for exposed histidyl residues in proteins [170]. However, this reagent also reacts with tyrosine, lysine, and cysteine residues. Under certain conditions it has been used for determining both the total tyrosine and histidine contents of proteins and their preferential reactivities.

2.6 Indoles

Tryptophan contains the largest aromatic side chain found in proteins. Its frequency is the lowest among all the amino acids, and although it has been suggested that tryptophan might participate in the catalytic action of some enzymes, its role is presumably limited in most cases to structural interactions such as hydrophobic bonding (Table 1). The indole ring of tryptophan is extremely susceptible to irreversible oxidation, yielding multiple complex products. This has greatly hindered progress not only with regard to the protein chemistry of tryptophan residues, but also in the development of accurate analytical procedures (see Section 3.1). The indole moiety of tryptophan is readily modified by alkylation, formylation, oxidation, and ozonolysis (Table 9) [18, 19]. In this section we are primarily concerned with electrophilic substitution reactions of the indole ring in proteins. These modifications occur chiefly at ring position 3, the alkyl-substituted carbon atom.

Reactive benzyl halides are selective alkylating agents for tryptophan residues in proteins [191]. The reagents most often used for this purpose are 2-hydroxy-5-nitrobenzyl bromide (Koshland's reagent) and the analogous dimethyl-(2-hydroxy-5-nitrobenzyl) sulfonium salts which have a higher solubility in aqueous solution. These benzyl halides are reactive as electrophiles because of the resonance stabilization of the incipient carbonium ion by the ortho hydroxyl substituent (Scheme 10). In the absence of sulfhydryl groups, the reagent appears to be quite specific for tryptophan under appropriate conditions of pH, since the ortho hydroxyl renders the methionine adduct extremely unstable. The reaction is complex, and the formation of multiple products detracts from the utility of these

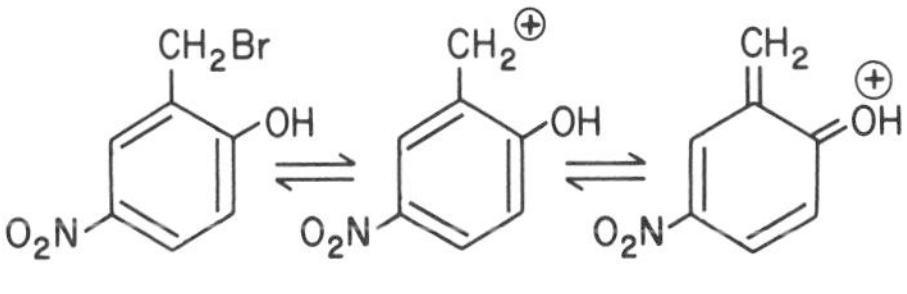

Scheme 10

compounds as analytical reagents [192]. As many as 10 different products were detected in the reaction of *N*-acetyl-L-tryptophan amide with dimethyl-(2-hydroxy-5-nitrobenzyl) sulfonium bromide [193]. Four of the products were characterized and are shown in Scheme 11.

Although the group specificity of sulfenyl halides has the same limitation as that of the benzyl halides, the former reagents yield a single product, namely, the derivative with a thioether function at the 2-position of the indole nucleus [194]. The reaction can be

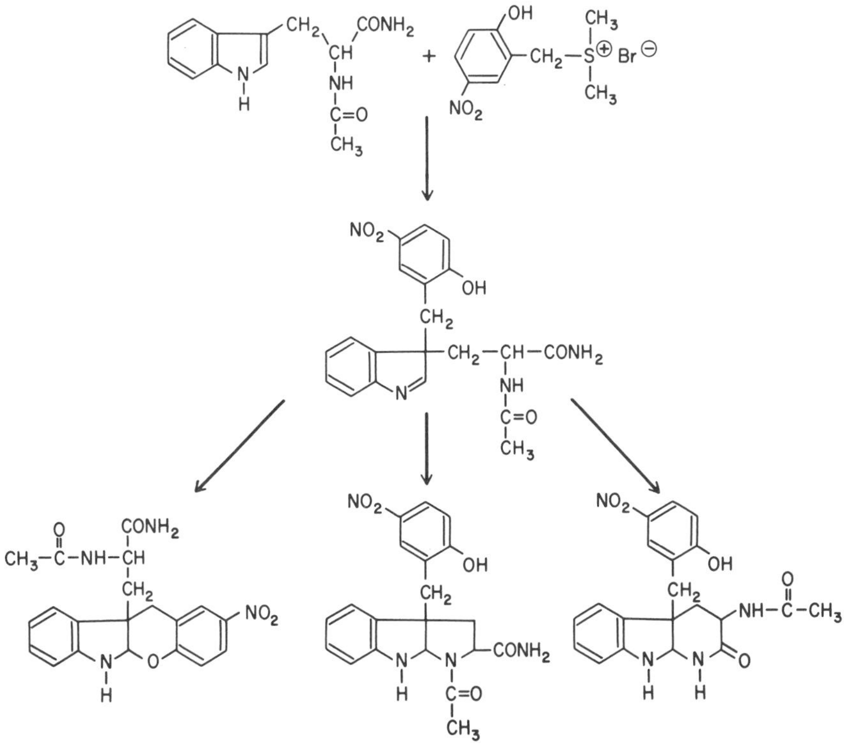

Scheme 11

TABLE 9 CHEMICAL MODIFICATION OF INDOLE GROUPS

Reagent	Protein	Results	Reference
2-Hydroxy-5-nitrobenzyl bromide	Malate dehydrogenase	Complex alkylation; 1 tryptophan residue per subunit modified in 8 M urea; inactivation	196
	Cobrotoxin	Single tryptophan modified; lowered toxicity	197
	Ribonuclease T$_1$	Trp-59 modified in 8 M urea	198, 199
	Erabutoxin a	Trp-29 alkylated; loss of toxicity	200
	Lysozyme	Trp-62, Trp-63, Trp-108, Trp-123 alkylated; Trp-28 reacts slowly; inactivation	201
	Myelin encephalitogenic basic protein	Single tryptophan alkylated; loss of neurological activity	202
	Bovine pancreatic deoxyribonuclease	2 of 3 tryptophan residues reacted in native with 25% loss of activity	203
	Porcine pituitary hormone	Single tryptophan alkylated; inactivation	204
	Aspergillus saitoi ribonuclease	Tryptophans reacted in 6 M guanidine	205
	Neuraminidase	Tryptophan alkylated; inactivation; substrate protected	206
	Bovine trypsinogen	Trp-199 modified with ability of zymogen to activate lost	207
	Transferrin	Tryptophan modified; metal binding decreased	208

Reagent	Protein	Effect	Ref.
Dimethyl-(2-hydroxy-5-nitrobenzyl) sulfonium salts	Transferrin	Tryptophan alkylated; no effect on metal binding	208
	Yeast transketolase	3 of 18 tryptophan residues modified; K_m decreased	209
	Bovine α-lactalbumin	Trp-26, Trp-104, Trp-118 reacted; 1 residue unreactive	210, 211
	Bovine chymotrypsinogen A	Trp-215 modified; zymogen unable to activate	212
	Carbonic anhydrase	1 of 6 tryptophan residues alkylated	213
2-Nitrophenylsulfenyl chloride	Myelin encephalitogenic basic protein	Single tryptophan converted to 2-thioether derivative; loss of neurological activity	202
	Staphylococcal nuclease	Single tryptophan modified with 50% loss of activity	214
	Transferrin	Single tryptophan modified; metal binding decreased	208
	Lysozyme	Trp-62 modified; inactivation	215
	Cytochrome c	Single tryptophan converted to N-formyl derivative; inactivation	216
Formic acid-HCl	Thioredoxin	2 tryptophans modified; inactivation	217
	Myelin encephalitogenic basic protein	Single tryptophan modified; loss of neurological activity	202

carried out in acidic medium where protonation of the amino groups prevents them from reacting. As in the case of the benzyl halides, free cysteines also react, yielding mixed disulfides from which the thiol function may be readily recovered by mild reductive procedures. When the reagent contains a nitrophenyl group, introduction of this chromophore allows one to use the reaction for quantitative determination of accessible tryptophans.

Reaction of tryptophan with hydrochloric acid dissolved in formic acid leads to formation of N-formyltryptophan [195]. The modification is reversible at pH 9, and no reaction occurs at amino or hydroxyl groups. Thus formylation would be highly selective for tryptophan, but the reaction conditions limit its use to proteins in which the preservation of the native structure is not required.

2.7 Phenols

Tyrosine contributes significantly to the ultraviolet spectra of proteins in the region of 280 nm. The phenol ring of tyrosine residues has an intrinsic pK_a = 9.6, and its spectral properties provide a convenient method for quantitation of tyrosines. By this approach, it has been found that a large number of phenolic hydrogens are resistant to ionization in the native protein, and that high pH or denaturation is necessary for the production of the phenolate ion. In spite of this "buriedness" of the tyrosyl side chains, they readily undergo typical electrophilic substitution reactions of phenols, provided that small reagents are used which can gain access to the shielded aromatic ring (Table 10).

Tetranitromethane is a group-specific reagent widely employed for the nitration of tyrosyl residues in proteins [218]. Nitration is believed to result from the coupling of the nitrite free radical ($NO_2 \cdot$) with the aromatic ring, yielding mono- and dinitrophenols substituted in the 3- and 5-positions. Because of the free-radical nature of the reaction, the selectivity of the modification is somewhat less than optimum, even under mild conditions. Thus thiol groups also react to give sulfinic acids and even disulfides. Under more drastic conditions, that is, high pH and excess reagent, histidine, tryptophan, and methionine may also react, and cross-linking of tyrosine could give polymerization products [219–221]. Moreover, if the reaction is terminated by acidification, the resulting nitrous acid reacts with amino groups and causes deamination. Rigorous adherence to experimental procedure is essential if the goal is to obtain mononitrated tyrosine derivatives.

The phenolic hydroxyl of tyrosine is converted to the *O*-acetyl derivative at pH 7.5 by treatment with acetic anhydride or *N*-acetylimidazole [222]. Even at this moderate pH, both compounds react with thiol and amino groups, although *N*-acetylimidazole is somewhat more selective for tyrosine. The phenolic function can be regenerated with hydroxylamine at pH 7.5, or by short exposure to alkaline conditions. Dicarboxylic anhydrides also acylate the phenolic oxygen, but the resulting esters undergo rapid hydrolysis even at neutral pH [223].

Cyanuric fluoride (23, Scheme 12) acts in a manner similar to that of a reactive acyl halide in the modification of tyrosines. The

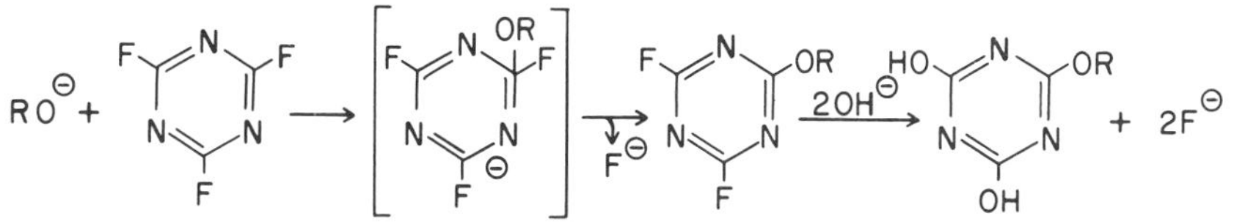

23

Scheme 12

reaction proceeds via nucleophilic attack of the phenoxide ion at a carbon atom of the heterocyclic ring [224, 225]. The substitution is carried out at pH 10, a condition that might promote undesirable conformational changes in many proteins. The relatively bulky reagent does not react with all tyrosines in proteins, perhaps reflecting its sensitivity to the degree of exposure of the individual tyrosines.

Tyrosine residues have been converted to the mono- and diiodo derivatives by a variety of iodinating reagents, although in all instances the reactive iodinating agent appears to be hypoiodous acid, HOI [226]. Monoiodohistidine may result in some cases as a side product.

2.8 Sulfhydryl Groups and Disulfides

The most labile functional group in proteins is the cysteine thiol. Sulfhydryl groups are uniquely reactive toward heavy metals because of the polarizability of the outer-shell electrons of the sulfur atom. Moreover, these thiol groups in proteins participate in almost all the reactions characteristic of the other functional side-chain sub-

TABLE 10 CHEMICAL MODIFICATION OF PHENOL GROUPS

Reagent	Protein	Results	Reference
Tetranitromethane	Ribonuclease	Mono-, di-, trinitrotyrosine and cross-linked products	221
	Trypsin	5 of 10 tyrosines nitrated	227
	Insulin	2 tyrosines converted to 3-nitrotyrosine and 2 cross-linked	228
	Arginine kinase	Single tyrosine nitrated; inactivation	229
	Tobacco mosaic virus antibodies	1 tyrosine modified per binding site	230
	Bacillus subtilis neutral protease	V_{max} decreased	231
	α-Amylase	1 tyrosine modified; inactivation	232
	Glucose-6-phosphate dehydrogenase	2 of 34 tyrosines modified; inactivation	233
	Hemerythrin	All 5 tyrosines reacted in apoprotein; 3 reacted in oxyhemerythrin	234
	Aspartate aminotransferase	No inactivation unless substrate present	235
	Lactogenic hormone	All 7 tyrosines reacted; full activity	236
	Cytochrome c	Tyr-67 nitrated; inactivation	237, 238
	Serum albumin	50% of tyrosines modified	239
	Thyroglobulin	90 of 125 tyrosines modified	239
	Hemoglobin	Tyrosines nitrated, 20% cross-linked	240
	Ribonuclease	Tyrosines nitrated, cross-linked	240
	Carbonic anhydrase B	3 of 8 tyrosines reacted; Tyr-20 reacts fastest	241
	Macroglobulin	Binding sites inactivated; ligand protected	242

Protein	Observation	Ref.
Thrombin	5 tyrosines modified; protease activity decreased; esterase activity increased	243
Insulin	40% mono-, 30% di-, and 18% trinitrotyrosine; reactivity: $A_{14} > A_{19} > B_{16} > B_{26}$	244
Lysozyme	Tyr-20 and Tyr-23 modified in native; Tyr-53 also modified in 8 M urea	245
Deoxyribonuclease	Single tyrosine modified; Ca^{2+} binding site destroyed	246
Glutamate dehydrogenase	Tyr-412 modified fastest	247
Yeast invertase	No effect on activity	248
Cobrotoxin	Tyr-35 modified with no loss of activity; Tyr-25 modified and activity lost	249
Stem bromelain	8 or 9 tyrosines modified; inactivation	250
α-Lactalbumin	Tyrosine and tryptophan modified; inactivation	251
Alkaline phosphatase	6 tyrosines reacted; active	252
Tropocollagen	All tyrosines modified; only 30% 3-nitrotyrosine produced	253
L-Asparaginase	Partial inactivation; 30% 3-nitrotyrosine; 60% cross-linked	254
Carboxypeptidase B	Tyr-248 modified; 30% activity	255
Pituitary-hormone	Tyrosines modified	256
Cell-stimulating hormone	5 of 7 tyrosines reacted; Tyr-21 and Tyr-59 did not	257
Aspartate transaminase	Inactivation	258
Sea snake neurotoxin	Inactivation	259
L-Asparaginase	Tyrosine modification caused aggregation	260
Naja neurotoxin	Tyr-24 modified; inactivation	261

TABLE 10 CHEMICAL MODIFICATION OF PHENOL GROUPS (*Continued*)

Reagent	Protein	Results	Reference
N-acetylimidazole	Trypsin and Kunitz Inhibitor	4 of 8 tyrosines protected in complex	262
	Bacillus subtilis α-amylase	Inactivation; reactivation with hydroxylamine	232
	Glucose-6-phosphate dehydrogenase	2 of 34 tyrosines acetylated; inactivation	233
	Glutamine synthetase	Tyrosines modified; partial inactivation	263
	Bence-Jones protein	2 of 4 tyrosines reacted	264
	Stem bromelain	8 or 9 tyrosines modified; full activity	250
	Hemerythrin	Metal protected 2 of 5 tyrosines; lysine also reacted	265
	Cell wall protein A	All 4 tyrosines modified; lysine also reacted	266
	Thermolysin	17 tyrosines acetylated; lysines also reacted	75
	Lysozyme	2 of 3 tyrosines modified; 4 of 7 lysines reacted	267
	Thrombin	4–5 tyrosines reacted; clotting activity decreased; benzamidine partially protected	268
Cyanuric fluoride	Chymotrypsin	3 of 4 tyrosines modified	269
	Apomyoglobin	2 of 3 tyrosines cyanurated	270
	Serum albumin	13 of 19 tyrosines reacted	270
	Carbonic anhydrase	6 of 8 tyrosines modified	270
	Elastase	8 of 10 tyrosines reacted	271

Iodination (hypoiodous acid)	Glyceraldehyde-3-phosphate dehydrogenase	Reactivity: Tyr-46 > Tyr-39 and Tyr-42	272
	Bence-Jones protein	Tyr-173 reacted	273, 274
	Glyceraldehyde-3-phosphate dehydrogenase	6 of 9 tyrosine residues reacted in native to diiodo derivatives	275, 276
	Bacillus subtilis neutral protease	Tyrosines modified; activity with large substrates decreased and small substrates increased	231
	Ferricytochrome c	Tyr-67 and Tyr-74 iodinated with partial inactivation	277
	Penicillinase	Tyrosine and tryptophan reacted; inactivation	278
	Carboxypeptidase B	Substrate protects Thr-Ile-Tyr-Pro-Ala	279
	Arginine kinase	1 tyrosine iodinated; inactivation	280
	Insulin	Tyr-A_{19} and Tyr-B_{16} modified and inactivation	281
	Ceruloplasmin	Tyrosine and histidines modified	282
	Scorpion neurotoxin	Tyr-8 reacted in native; Tyr-5 and Tyr-14 reacted with inactivation	283
	Cytochrome b_5	Reactivity: Tyr-6 and Tyr-7 > Tyr-30 ≫ Tyr-27	284
Diazonium-1-*H*-tetrazole	Hemoglobin	8 of 12 tyrosines reacted; histidine also modified	285
	Myoglobin	Both tyrosines modified; histidine also reacted	285
	RNA polymerase	33 of 100 tyrosines modified; histidine also reacted	189

stituents, namely, alkylation, acylation, addition, and oxidation-reduction reactions. Sulfhydryl groups in proteins are most reactive at slightly alkaline pH values, as a consequence of their intrinsic ionization constant of about 9 (Table 2). Relative to the other amino acid constituents of proteins, cysteine and its oxidized form, cystine, occur less frequently on the average (Table 1). The accessibility factor in Table 1 for the cysteine side chain is misleading, since it does not distinguish between reduced and oxidized forms of the sulfhydryl group. Disulfide bonds are often found in the interior of proteins, whereas free sulfhydryl groups are hydrophilic, are usually exposed to solvent, and often play a central role in the function of the protein in question.

The high reactivity of thiol groups at neutral pH simplifies the search for reagents that may be employed for their specific modification. Compounds such as haloacetates, haloacetamides, and phenacyl halides are examples of reagents that alkylate sulfhydryl groups via displacement of the halogen by the nucleophilic sulfide ion [168, 286]. The electrophile N-ethylmaleimide alkylates thiol groups by addition of sulfur at the double bond [287]. Numerous reagents of similar chemistry have been described in earlier reviews of sulfhydryl group modifications (cf. [20, 25]). For the purpose of this chapter, extensive documentation of these reactions would be largely redundant. Table 11 summarizes the results of some recent applications of thiol modifications in a number of proteins. From the chemist's point of view, however, perhaps the most interesting aspects of protein sulfhydryl chemistry relate to modifications that facilitate analysis of cysteine and cysteine phenylthiohydantoin (PTH) and those that promote polypetide chain scission at cysteinyl residues. Several modifications have been useful in providing good quantitation of cysteine in acid hydrolyzates of proteins, and many of these are discussed in Section 3.1.1. The Ellman reagent [5,5'-dithiobis(2-nitrobenzoic acid), DTNB] has been and continues to be a useful means of determining cysteine content in native and denatured proteins [288]. This reagent modifies thiol groups by an exchange reaction to form the mixed disulfide of the protein with the concomitant release of the 2-nitro-5-thiobenzoate ion, the absorbance of which provides a means of quantitating spectro-photometrically the extent of substitution. A thorough account of the use of this reagent is given by Glazer [25].

Conversion of cysteinyl residues to the S-methyl [166, 289] and S-pyridylethyl [290] derivatives has been useful both in quantitating cysteine in acid hydrolyzates and in identifying the corresponding

PTHs during the course of sequence analysis [291, 474]. Two procedures for S-methylation have been reported recently, which are highly specific for thiol groups in denatured proteins and peptides. That reported by Heinrikson [166] involves partial or complete S-alkylation of the native or the reduced and denatured protein with methyl-p-nitrobenzenesulfonate. An alternate approach, limited to studies in which the complete derivatization of cysteine and cystine is desired, is that described by Meinenhofer et al. [289], in which the protein reduced by dithiothreitol is alkylated with methyl chloride in liquid ammonia. The S-methylated proteins obtained by these methods have provided useful derivatives for the study of cyanogen bromide-mediated peptide chain cleavage at these sites (cf. Section 3.2.2.1). Conversion of cysteine to the S-cyano derivative by reaction with 2-nitro-5-thiocyanatobenzoate provides a means for the reversible blockage of sulfhydryl groups, and a radiolabel may be easily introduced by this means [292]. Moreover, as described in Section 3.2.2.1, the S-cyanylated protein derivatives may, under appropriate conditions, undergo selective chain cleavage in high yield at the modified sites [293].

Thus, because of the unique reactivity of the sulfhydryl group, selective modification of cysteines in proteins can be achieved by a variety of reagents and reaction types. Labeling, quantitation, oxidation-reduction, and alkylation of cysteine residues are straightforward procedures and yield predictable results for most proteins. A general technique is still lacking, however, for the conversion of a cysteine side chain into a serine side chain (cf. Section 3.2.2.1), or for a well-controlled quantitative conversion into the sulfenic, sulfinic, and sulfonic acid forms.

2.9 Thioethers

The long alkyl side chain of methionine renders it rather hydrophobic, and its position in the tertiary structure of proteins is such that it is usually shielded from solvent (Table 1). This relatively infrequently occurring amino acid has never been clearly implicated to display any catalytic role in enzymes (cf., however, [346, 347]) in spite of the fact that the sulfur of the thioether group is a potent nucleophile and it readily forms sulfonium salts with alkylating agents such as alkyl and benzyl halides (Table 12) [168]. Even in the case of methyl transfer enzymes, the functional methionine is attached to a nucleotide rather than to the protein that carries out the catalytic function. This paradox of protein chemistry might

TABLE 11   CHEMICAL MODIFICATION OF THIOL GROUPS

Reagent	Protein	Results	Reference
Methyl-*p*-nitrobenzene-sulfonate	Insulin B chain	S-methylation of 2 cysteine residues	166
	Rhodanese	2 cysteines/monomer methylated	166
	Papain	Active-site cysteine reacts only in 7 *M* urea	166
(2-Bromoethyl)trimethyl-ammonium bromide	Insulin	Cysteine converted to trimethylaminoethyl-cysteine	294
Ethylenimine in liquid ammonia	Insulin, basic trypsin inhibitor, lysozyme	Quantitative aminoethylation of thiol groups	295
p-Nitrostyrene	Serum albumin, lysozyme, Immunoglobulin G	Quantitative nitrophenylethylation of thiol groups	296
4-Vinylpyridine	Bovine serum albumin, β-lactoglobulin, lysozyme, ribonuclease, ovalbumin	Quantitative pyridylethylation of thiol groups	290
Iodoacetate	Glyceraldehyde-3-phosphate dehydrogenase	1 cysteine per monomer alkylated; inactivation	297
	Aldolase	3 cysteines per subunit alkylated; substrate protected 1 cysteine	298
	Liver alcohol dehydrogenase	Cysteine alkylated; 3% activity	299
	High-density lipoprotein	Cysteine alkylated; conformational change	300
	Trypsinogen	Disulfide 179–203 alkylated	301–303
	Actin	NH$_2$-terminal thiol alkylated	304
Iodoacetamide	Malate dehydrogenase	2 cysteines modified; inactivation; NADH protected	305
	Creatine kinase	Essential thiol group alkylated	306
	Hemoglobin	Cys-93β alkylated	307–309

Reagent	Enzyme	Description	Reference
Bromoacetate	Yeast aldolase	Essential cysteine modified; K_m for NADH decreased slightly	310
Bromoacetate	Muscle aldolase	3 of 7 cysteines carboxymethylated; 1 protected by phosphate	311
Bromopyruvate	Aspartate aminotransferase	Inactivation; cysteine modified	312
	Ficin	Cysteine cross-linked with histidine	313
1,3-Dibromoacetone	Stem bromelain	Cysteine cross-linked with histidine	314
	Glyceraldehyde-3-phosphate dehydrogenase	1-cysteine alkylated	315, 316
Bromoacetol phosphate	Aldolase	Essential cysteine alkylated	317
Maleimide	Lactate dehydrogenase	Essential cysteine alkylated; NADH protected	318
	Arginine kinase	Essential cysteine alkylated	319
	Hemoglobin	α and β subunits alkylated	320–322
	Malate dehydrogenase	Essential cysteine modified; NADH protected	323
	Tryptophan synthetase	Cys-80 and Cys-117 reacted in α subunit	324, 325
	Hemoglobin	Cross-linking with bifunctional maleimides; Cys-93β with His-97β; Cys-93β with Val-1β	326, 327
	Phosphoenolpyruvate decarboxylase	Essential cysteine alkylated	328
β-Lactoglobulin		Single cysteine alkylated and conformation altered	329
	Succinic dehydrogenase	Several cysteines alkylated; one protected by flavin	330
N-Tosyl-L-phenylalanine chloromethyl ketone	Luciferase	Essential cysteine modified	331
N-Tosyl-L-lysine chloromethyl ketone	Clostripain	Essential cysteine alkylated	332

TABLE 11 CHEMICAL MODIFICATION OF THIOL GROUPS (*Continued*)

Reagent	Protein	Results	Reference
α-Bromo-4-hydroxy-3-nitroacetophenone	Papain	Cys-25 alkylated; inactivation	286
5,5'-Dithiobis(2-nitrobenzoic acid)	Isocitrate dehydrogenase	2 essential cysteins reacted; both protected by isocitrate	333
	Glyceraldehyde-3-phosphate dehydrogenase	1 of 5 cysteines per subunit reacted rapidly with inactivation	334
	Phosphorylase b	Several cysteines reacted; inactivation and dissociation into subunits	335
	Lipase	1 of 2 cysteines reacted in native	336
	Aspartate transcarbamylase	1 cysteine per subunit modified with inactivation	337
	Aspartate aminotransferase	2 sulfhydryl groups oxidized; inactivation	338
	Cystathionase	12 sulfhydryl groups reacted in native; 20 reacted in 8 M urea	339
	Pyrocatechase	Cysteine in apoprotein reacted; holoprotein unreactive	340
	Papain	Essential cysteine modified	341
2-Mercaptopyridine	Phosphofructokinase	Cysteine modified and still active	342
	Immunoglobulin M	Disulfides interchanged	343
	Bromelain	Essential cysteine modified	344
Butylisocyanate	Yeast alcohol dehydrogenase	Three cysteines reacted; Cys-Ala-Gly-Ile-Thr-Ala	345

TABLE 12 CHEMICAL MODIFICATION OF THIOETHER GROUPS

Reagent	Protein	Results	Reference
Iodoacetic acid	Staphylococcal enterotoxin B	4 of 8 methionine residues alkylated to carboxymethylmethionine sulfonium salts with inactivation	350
	Isocitrate dehydrogenase	One methionine alkylated and inactivation	351
Bromoacetic acid	Ferricytochrome c	Met-65 and Met-80 alkylated	352
	Ferricytochrome c	Met-80 alkylated	353
Trichloromethane sulfonyl chloride	α-Chymotrypsin	Met-192 oxidized to sulfoxide and active conformation stabilized	349

somehow be related to the fact that small molecules are transported more easily within the cell than are large proteins. One methionine residue (Met-192) is close to the active site of chymotrypsin, but its role is limited to forming part of the hydrophobic binding pocket [348].

Selective modification of methionine can be achieved because rates of alkylation or oxidation are virtually pH independent, whereas reactions of other functional groups are inhibited by decreasing pH. For this reason, absolute specificity for the modification of methionine is possible and has often been achieved, for example, in the case of chymotrypsin. Most recently, Taylor et al. [349] oxidized Met-192 to the sulfoxide with trichloromethanesulfonyl chloride. This reagent is a source of positive halogen which is the oxidizing agent (Scheme 13).

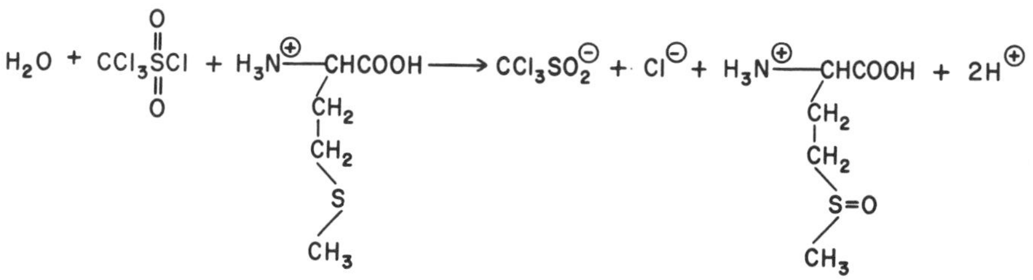

$$H_2O + CCl_3\overset{\overset{O}{\|}}{\underset{\underset{O}{\|}}{S}}Cl + H_3\overset{\oplus}{N}\text{---}CHCOOH \longrightarrow CCl_3SO_2^{\ominus} + Cl^{\ominus} + H_3\overset{\oplus}{N}\text{---}CHCOOH + 2H^{\oplus}$$

Scheme 13

3 RECENT ADVANCES IN PROTEIN COVALENT STRUCTURAL ANALYSIS

In the past, knowledge of the purity, size, and amino acid composition of proteins and peptides was a basic prerequisite to the undertaking of covalent structural analysis of these substances. With the development and increasing application in recent years of automated equipment for performing the Edman degradation, the strategy of sequence determination has changed in various ways. In contrast to the conventional approach, which utilizes fragmentation methods designed to produce small peptides amenable to manual sequence analysis, the automated Edman procedure originally described by Edman and Begg [354] is best suited to the degradation of larger fragments and intact proteins. This has led to increased interest in specific chemical cleavage procedures by means of which polypeptide chain scission is limited to sites occupied by minor amino acid constituents of proteins such as, for example, methionine and tryptophan. Moreover, in view of the higher repetitive efficiency of the automated Edman procedure and the resulting increase in reliability of quantitative as well as qualitative evaluation at each step, the stringent requirements for sample purity have been overcome to a degree, and in many instances mixtures of peptides may be sequenced simultaneously. This section is devoted to a discussion of advances in amino acid analysis, specific chemical cleavage procedures, and methods for primary structural analysis that have been particularly relevant to the field of protein chemistry in recent years.

3.1 Quantitative Determination of Amino Acid Composition

The impact on contemporary structural biochemistry derived from the intelligent application of the chromatographic method has probably never been more spectacularly realized than in the development of automated ion-exchange chromatography for amino acid analysis. Although in comparison with the prototype analyzer, modern instruments have tremendously increased capabilities in terms of speed, sensitivity, flexibility, and data analysis, the basic chemistry pioneered in the late 1950s by Moore and Stein and their collaborators at the Rockefeller University remains essentially unchanged [9, 10]. One needs only to consider that the time required for the quantitative amino acid analysis of a protein hydrolyzate has been reduced from a matter of years with poor accuracy to about an hour with high precision, to realize how much our rapidly expanding knowledge of protein structure has depended upon this method. Two recent articles by Moore [355] and by Moore and Stein [356] deal with the subject in considerable detail; only some of the more salient points covered by these publications are mentioned here.

3.1.1 **Hydrolysis of Proteins** Compared to the dramatic increases in speed and sensitivity that have attended the continuous development and refinement of the analytical procedure itself, relatively few changes have occurred in the past 15 years with respect to the hydrolytic step in the analysis. Ideally, one would wish to have a procedure that would, in a matter of minutes, quantitatively hydrolyze a protein or peptide to its constituent amino acids without destruction of the same.

At the present time, however, fulfillment of this ideal seems an exceedingly remote possibility. Roach and Gehrke [357] have written a review on the hydrolysis of proteins in which is described a 4-hr hydrolysis in 6 N HCl at 145°C. Nevertheless, the hydrolytic procedures detailed earlier by Moore and Stein [11], namely, incubation of protein solutions in 6 N HCl in evacuated Pyrex tubes for 22-100 hr at 110°C, remain the most commonly employed means of preparing acid hydrolyzates for the amino acid analyzer. Analysis of samples hydrolyzed for varying lengths of time provides data for back extrapolation to zero time to correct for destructive losses of serine, threonine, and tyrosine; the longest time of hydrolysis usually gives the best estimation for amino acids with bulky hydrophobic side chains, which are liberated more slowly during the hydrolysis. Addition of scavengers such as phenol [358, 359] ensures higher recoveries of tyrosine, but in any case most of the amino acids may be accurately estimated by this rather lengthy hydrolytic procedure. Two exceptions are cysteine (or cystine) and tryptophan.

Proper evaluation of the cysteine plus cystine content of a protein requires that these amino acids be converted in high yield to derivatives that are stable to the hydrolytic conditions or, after hydrolysis, to derivatives that may be quantitated precisely. Performic acid oxidation [360] converts these residues to cysteic acid which may be accurately estimated on an analyzer, but this method suffers from the disadvantage that methionine and tryptophan residues, potential sites of specific chemical cleavage, are also oxidized. Milder conditions, under which these deleterious side reactions are avoided, usually involve reduction of the disulfide bonds, followed by alkylation to yield an S-alkylcysteine derivative which is both stable to acid hydrolysis and readily distinguishable among the other amino acid components of proteins under the routine conditions of amino acid analysis. Hydrolyzates of derivatives S-alkylated with haloacetate or haloacetamide contain carboxymethylcysteine which appears just ahead of aspartic acid in the elution profile [361]. Methylcysteine [362, 166] is eluted on the trailing edge of proline, and aminoethyl- [363] and pyridylethyl- [290] cysteine may be distinguished among the normal basic amino acid constituents of proteins. Inglis and Liu [364] have published a method for the estimation of cysteine and cystine in acid hydrolyzates by reduction with dithiothreitol followed by reaction with tetrathionate which quantitatively converts these residues to S-sulfocysteine. All the procedures outlined above, alone or in combination, serve to yield accurate estimations of the cysteine and cystine content of the protein in question. For structural work, the means of dealing with these residues is largely dictated by the strategy to be employed in subsequent steps of the sequence analysis. Performic acid-oxidized protein derivatives are usually quite soluble and amenable to proteolytic digestion at neutral pH, whereas the solubility of reduced and alkylated proteins may vary considerably depending upon the nature of the modifying adjunct. Cysteic acid residues, however, are not easily placed in a sequence when direct Edman degradative methods are employed. Further considerations with regard to the possible introduction of new sites for tryptic cleavage and to the particular means employed for identifying the position of the cysteinyl derivative within a peptide sequence are discussed in Sections 3.2.1, 3.2.2, and 3.3.1.

The remaining amino acid that requires special attention because of degradative losses during hydrolysis in HCl is tryptophan. In the past, this amino acid has been determined spectrophotometrically [365], colorimetrically [366, 367], by alkaline hydrolysis [368] or, more recently, by hydrolysis in 6 N HCl containing 4% thioglycolic acid [369]. The effect of scavengers on increasing yields of tryptophan during hydrolysis in HCl has been studied by Gruen and Nicholls [370]. Two new developments in the quantitative estimation of tryptophan in proteins by hydrolysis and amino acid analysis have been of special importance to protein chemistry. The first, reported recently by Liu and Chang [371], involves hydrolysis of the protein in 3 N p-toluenesulfonic acid containing 0.2% 3-(2-amino-ethyl)indole (tryptamine) under the same conditions normally employed for this purpose. By eliminating HCl from the hydrolytic procedure, excellent recoveries of all the amino acids, including tryptophan, may be realized, and the analyses may be performed by conventional means. Exploring the method further, Liu replaced 3 N p-toluenesulfonic acid with 4 N methanesulfonic acid and, at a temperature of 125°C, hydrolysis for 20 hr is sufficient to liberate most of the resistant linkages in proteins. Under these conditions, the yield of tryptophan is quantitative, and the recoveries of threonine and serine are higher than in HCl.

We have made preliminary studies of protein hydrolysis with the extremely powerful and nonoxidizing trifluoromethanesulfonic acid. Under the same conditions employed with methanesulfonic acid, no advantages were observed with the trifluoro derivative. In spite of the lower boiling point of the latter compound, it was not easily removed under vacuum and, in view of its high toxicity and the precautions required in handling, it appears that methanesulfonic acid is the best choice for this purpose.

The obvious advantage of the Liu and Chang procedure is that the complete amino acid composition of a protein may be determined with a single analysis. Its greatest single drawback is that recoveries of tryptophan are not quantitative with glycoproteins, the carbohydrate content of which exceeds a few percent. Moreover, since the sulfonic acids are not volatile, a neutralization step is required prior to addition of the hydrolyzate to the amino acid analyzer. This dilution step could be significant in studies of proteins that are available in only minute quantities. Nevertheless, the use of sulfonic acids is an important step forward in the hydrolysis of proteins, and will undoubtedly be employed with increasing frequency in the analysis of proteins and peptides lacking carbohydrates.

The problem of carbohydrate with respect to tryptophan analysis has been overcome in the second procedure to be discussed; in fact, starch is employed as an antioxidant additive. This method, described by Hugli and Moore [372], is designed specifically for analysis of tryptophan and involves hydrolysis in 4.2 N NaOH containing 25 mg of starch per 0.6 ml of solution. Under these conditions, several amino acids are largely destroyed, but the recovery of tryptophan is strictly quantitative, somewhat better perhaps than that observed after hydrolysis with sulfonic acid. The advantage of the method is that it gives rigorously accurate tryptophan compositions in all proteins including those containing significant amounts of carbohydrate, and thus may be of greatest importance in determining the composition of this amino acid in a new protein. Drawbacks to the procedure relate primarily to the analytical manipulations. Hydrolysis must be performed in polyethylene liners inserted in Pyrex tubes which are then sealed *in vacuo*. Neutralization of the hydrolyzate is necessary prior to analysis. Perhaps the greatest disadvantage of the method is that it requires a special column for the separation of tryptophan from lysinoalanine which is formed in significant quantities during hydrolysis in base [373]. The fact that conventional analyzers cannot be employed for the determination, coupled with the necessity of a separate hydrolysis for the other amino acids, argues in favor of the methanesulfonic acid hydrolytic procedure for tryptophan in the majority of applications involving carbohydrate-free proteins and peptides. In any event, both approaches offer select advantages, and the combination truly provides the first sound analytical basis for the analysis of tryptophan.

3.1.2 **Amino Acid Analysis** Having considered some of the hydrolytic methods currently in vogue, it is appropriate to mention a few of the recent advances in contemporary commercial analyzers that have so drastically reduced the time and increased the sensitivity of amino acid analysis. The discussion is limited to ion-exchange procedures, without reference to the highly sensitive and rapid methods of gas chromatographic analysis. This preference is dictated not so much by a pessimism on our part toward the possibilities of the latter approach, as to the fact that gas chromatography has simply failed to provide as yet a challenging alternative to automated ion-exchange chromatography. It has always been assumed that the great disadvantage to gas chromatography, that is, the required derivatization step prior to analysis, would be overcome by the speed and sensitivity of the technique. In fact, however, even at a time when 5 hr were required for an analysis, the vast majority of protein chemists depended upon the ion-exchange analytical approach for the most reliable data both with respect to reproducibility and precision. Recent advances in ion-exchange technology have produced commercial analyzers with even increased precision, which rival the gas chromatographic method in regard to speed and sensitivity.

Of the various analyzers presently on the market, the Beckman Model 121M and the Durrum D-500 deserve special mention. Both utilize single-column modes of analysis, although the Beckman offers the standard two-column option originally described by Spackman, Stein, and Moore [10]. The analyzers are roughly comparable with respect to analysis time with column regeneration (1–2 hr), sensitivity (down to 100 pmoles), use of dimethyl sulfoxide as a solvent for ninhydrin [374], and price. However, whereas in the Beckman 121 M solvents are pumped in the conventional manner at relatively high pressures (700 psi), the Durrum D-500 offers an innovative hydraulic piston device which displaces eluting buffers, NaOH, and ninhydrin through various syringes at pressures of about 2500 psi. The single stainless-steel column of internal diameter 1.75 mm is packed with 8-μ beads of resin, and by employing fine-bore metal tubing throughout the system the sensitivity of the analysis is thereby greatly enhanced (cf. [375, 376]). Elution is performed with three buffers, the first two of which are the conventional pH 3.25 (A) and pH 4.25 (B) buffers [9, 10]; the latter (C) is a citrate buffer of higher ionic strength and pH. The standard procedure recommended by the Durrum Company utilizes a buffer C of pH 7.9, and histidine is eluted early in the analysis just after phenylalanine. Because of baseline shifts in this region of the chromatogram, Moore [377] has employed a buffer C of pH 6.5, which increases the total analysis time by a few minutes but which gives a more precise estimation of histidine. The effluent stream is monitored by a dual-beam photometric unit which scans at 590 mμ; a new photometer currently available provides a 440-mμ option for more sensitive detection of proline.

The most appealing aspect of the Durrum D-500 relates to the fact that it is fully computer operated. This offers advantages with respect to programming flexibility which are shared by most contemporary analyzers, but the paramount advantage resides in the data processing capacity of the Durrum D-500. By calibrating the instrument with a standard mixture of amino acids (usually 5 or 10 nmoles of each), a set of color constants is entered into the memory bank of the computer, each of which relates to a specific elution time, hence a specific amino acid. For each amino acid peak appearing on the chromatogram, the teletype prints out the elution time, name (in three-letter code), area, color factor for the particular

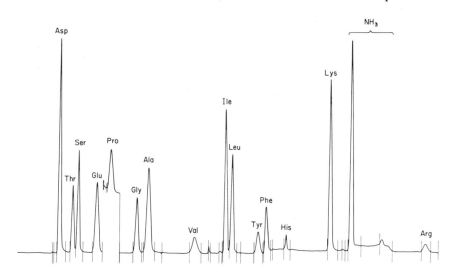

Figure 1   Elution profile obtained during the automated, ion-exchange amino acid analysis of a CNBr fragment from yeast inorganic pyrophosphatase on a Durrum D-500 analyzer [378]. Approximately 1.7 nmoles of peptide were subjected to analysis (cf. Fig. 2) on this accelerated, high-sensitivity analyzer.

ANALYSIS OF CNBR II - PPASE
PROC= 2   RUN= 1

PEAK NAME		MIN-SEC		TYPE	AREA	QUANT	FACTOR	BASE
1	ASP	13	32	Ø	47911	17.2	27787	1ØØ
2	THR	15	45	1	1517Ø	5.5	27765	1ØØ
3	SER	16	44	2	22285	7.6	294Ø7	1ØØ
4	GLU	2Ø	4	Ø	24779	8.7	2832Ø	1ØØ
5	PRO	22	31	Ø	7811	6.8	11551	1Ø76
6	GLY	27	5	Ø	17Ø4Ø	5.9	29Ø88	97
7	ALA	29	13	Ø	32556	11.2	3Ø753	97
8	VAL	37	34	Ø	931Ø	3.3	2817Ø	97
9	ILE	42	56	1	318Ø3	1Ø.4	3Ø443	97
1Ø	LEU	44	4	2	26318	7.8	33616	97
11	TYR	48	43	Ø	871Ø	3.5	24574	97
12	PHE	5Ø	8	Ø	13342	5.4	247Ø7	97
13	HIS	53	4Ø	Ø	5155	1.7	29617	97
14	LYS	61	24	Ø	41974	14.4	292Ø7	97
15	NH4	65	7	Ø	58244	2Ø.6	28255	97
16		7Ø	38	Ø	12375	4.6	27ØØØ	97
17	ARG	78	19	Ø	714Ø	2.9	245Ø1	97

```
        79   2Ø
ERR=   Ø
```

PEAK NAME		MIN-SEC		% CONC
1	ASP	13	32	12.34
2	THR	15	45	3.948
3	SER	16	44	5.455
4	GLU	2Ø	4	6.245
5	PRO	22	31	4.881
6	GLY	27	5	4.235
7	ALA	29	13	8.Ø4Ø
8	VAL	37	34	2.368
9	ILE	42	56	7.465
1Ø	LEU	44	4	5.599
11	TYR	48	43	2.512
12	PHE	5Ø	8	3.876
13	HIS	53	4Ø	1.22Ø
14	LYS	61	24	1Ø.33
15	NH4	65	7	14.78
16		7Ø	38	4.602
17	ARG	78	19	2.Ø81

Figure 2   Computer monitor of the amino acid analysis corresponding to the chromatogram depicted in Figure 1. The number of residues of each amino acid in the fragment may be calculated by dividing its "Quant" figure by 1.7.

amino acid, baseline value, and quantity in nanomoles. The last figure is derived by dividing the area by the color constant and multiplying by the number of nanomoles of sample employed in the calibration step. At the end of an analysis, the percent composition of each amino acid constituent is printed out together with the names and elution times. Figure 1 is the elution profile obtained from the amino acid analysis of a CNBr fragment from yeast inorganic pyrophosphatase [378]; the computer printout for this run is duplicated in Figure 2. This analysis required about 80 min. Individual peaks are very sharp and well separated. In this system, the effluent is monitored for absorbance at 590 nm, and since proline gives very low color yields at this wavelength, the sensitivity is automatically increased 10-fold during its elution (Figure 1). By such stringent monitoring of the whole process of integration, together with the actual chromatogram of the run, it is possible to obtain the most precise estimation of each amino acid in an unknown sample. Moore [355] has reported a maximum deviation of well under 1% with the Durrum D-500 analyzer.

With regard to the question of sensitivity, it is hard to imagine much further practical improvement without requiring extraordinary precautions against outside contamination of samples. One noteworthy attempt in this direction has been the application of a new reagent, fluorescamine (4-phenylspiro[furan-2(3H), 1'-phthalan]-3,3'-dione), to the detection and quantitative evaluation of peptides and amino acids [379, 380]. This nonfluorescent reagent, the tradename of which is Fluram, reacts almost instantaneously with primary amines at room temperature and pH 8.5–9.0 to yield highly fluorescent products (Scheme 14). The reagent is rapidly destroyed by a competitive reaction with water, but these products are nonfluorescent. An obvious application of this

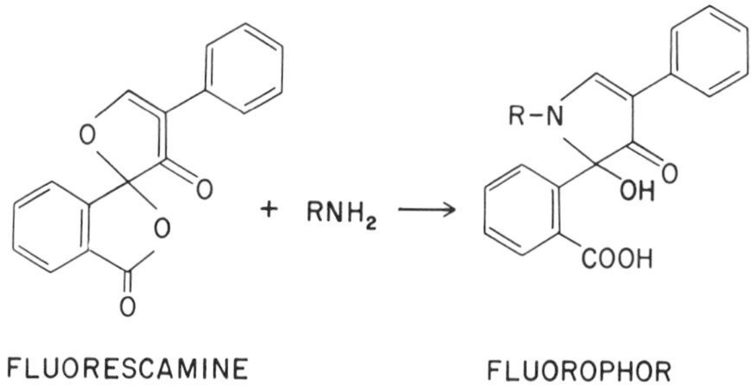

FLUORESCAMINE   FLUOROPHOR

Scheme 14

reagent would be in replacing ninhydrin as a means of detecting and quantitating amino acids in the effluent stream of an analyzer. By mixing the same with a solution of borate buffer at suitably high pH, followed by a solution of fluorescamine in acetone, fluorescent products would be generated without the high temperatures required in the ninhydrin reaction. In fact, a prototype of this kind has been developed that incorporates a simple recording fluorometer in place of the usual photometric detection unit [379]. For sake of comparison, two chromatograms, one derived from 10 nmoles of amino acid standard analyzed on a Durrum D-500 instrument, and the other obtained from 0.25 nmoles of sample on the fluorescamine analyzer, are shown side by side in Figure 3. Reaction of fluorescamine with ammonia yields derivatives which fluoresce orders of magnitude less than primary amines, hence ammonia does not interfere with analysis of the basic amino acid constituents of proteins, and many of the baseline shift problems caused by ammonia in buffers are obviated. These advantages are offset somewhat by the lack of reactivity with proline, a secondary amine. However, further work along these lines has provided the means by which proline may be converted to a reactive derivative [381] and thus quantitated along with the other amino acids. Clearly, this approach is in its infancy and further developments will doubtless follow in rapid succession as the methodology is applied more generally to the problems of protein and amino acid chemistry.

The foregoing discussion has dealt with methods for separation of L-amino acids; these methods do not distinguish between D- and L-enantiomorphs, and this could be disadvantageous in some circumstances. The stereochemical purity of amino acid residues in synthetic peptides, for example, has become of increasing interest in recent years, with the great expansion of activity in the field of peptide synthesis. Moreover, many antibiotic peptides contain D-amino acid residues. Procedures for separating the D- and L-antipodes of various amino acids on an amino acid analyzer have been devised by Manning and co-workers [382–384], and application of these methods to the problems mentioned above should greatly facilitate their solution.

In view of the staggering progress already achieved in shortening the time of analysis, pessimism with regard to how far one may go in speeding up the process seems unwarranted. At present, analyses are being run in about 1 hr on the Durrum D-500 analyzer. Coincidentally, automated Edman degradative procedures produce anilinothiazolinone derivatives of the NH_2-terminal amino acids as

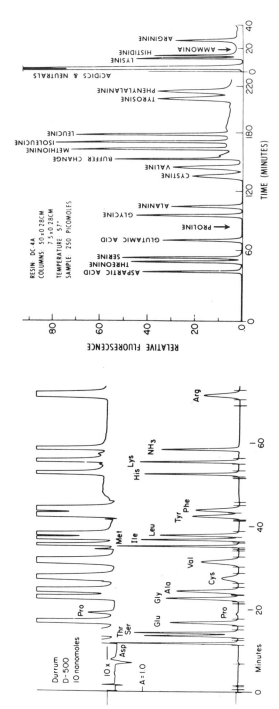

Figure 3 Comparison of sensitivities in automated ion-exchange amino acid analysis afforded by monitoring the effluent stream by reaction with ninhydrin (left) and fluorescamine (right). The chromatogram on the left was obtained by analysis of a standard mixture of amino acids (10 nmoles each) on an accelerated, single-column Durrum D-500 analyzer; the upper curve is a 10-fold expansion in sensitivity, which permits integration of proline. Data from Moore [355] by permission. The profile on the right shows separation on a standard two-column analyzer of a similar sample diluted 40-fold. In this case, the effluent stream is mixed with borate buffer at elevated pH and then reacted with fluorescamine. The fluorescent amino acids are quantitated by means of a recording microfluorometer. From Udenfriend et al. [379] by permission.

200

they are sequentially removed from the polypeptide at a rate of about one per hour. The analyzer thus provides a realistic means with respect to the time scale for providing quantitative estimations of the amino acids produced by hydrolytic back conversion of the phenylthiohydantoin (PTH) amino acids produced from these thiazolinones. This approach is discussed at greater length in Section 3.3.1, but high-speed analyzers have played a great part in making it a practical alternative, as well as a complement to, gas chromatographic analysis. As the time of analysis is reduced more and more, it is increasingly clear that automated means of integration and data processing are not just a luxury but a necessity for the precise and efficient estimation of amino acids. The advances made to date in this regard have been spectacular, and future developments in the field should provide instrumentation that will find ever widespread application in protein chemistry.

3.2 Specific Chemical and Enzymic Cleavage of Proteins

A necessary prelude to satisfy any expectation of quantitative protein cleavage, whether by proteases or by chemical means, requires that the polypeptide of interest be disrupted or converted to a random coil so that all pertinent peptide bonds are equally exposed to the cleaving enzyme or reagent. Specific chemical fragmentation is often performed in highly disruptive acidic media, or in the presence of protein denaturants such as 6 M guanidinium hydrochloride. Such conditions are usually too drastic for enzymic digestion, and very often selective chemical modification, for example, by reduction and alkylation or by acylation, provides the key for obtaining suitably denatured protein substrates.

3.2.1 **Limited Specific Cleavage by Enzymes** The enzymes commonly employed in the past for polypeptide chain scission, namely, trypsin, chymotrypsin, pepsin, papain, thermolysin, and so on, are still in widespread use, both in conventional sequence analysis and in conjunction with automated analytical procedures. With respect to the latter approach to sequence determination, enzymic cleavages restricted to peptide bonds adjacent to specific amino acids have been of great interest. One particularly fruitful example has been the tryptic cleavage of proteins at arginine residues after modification of the ϵ-amino groups of lysines by succinylation, maleylation, or citraconylation. In the case of acidic proteases that have few trypsin-susceptible sites of cleavage, reduction and aminoethylation of cysteinyl residues have been employed to generate bonds that may be hydrolyzed by trypsin [363, 385].

Two enzymes specific for a given amino acid residue have been described recently and may find useful application in structural analysis. One, an extracellular protease from *Staphylococcus aureus*, isolated by Houmard and Drapeau [386], appears to cleave peptide bonds specifically at the carboxyl terminal sides of glutamyl and aspartyl residues when reactions are performed at pH 7.8 in phosphate buffers. In ammonium bicarbonate (pH 7.8) or ammonium acetate (pH 4.0), however, cleavage is restricted to glutamic acid residues in the peptide chain. Hydrolysis by the proteinase of all nine different glutamyl bonds that occur in insulin, ribonuclease, and lysozyme point to a broad specificity regarding the chemical nature of neighboring amino acid side chains. This relative newcomer to the protein chemistry scene should be extremely valuable to structural studies. Furthermore, its small size (mol wt 12,000), the fact that it is extracellular and therefore readily available, and its sensitivity to the "serine enzyme" inhibitor, diisopropyl phosphorofluoridate, make the staphylococcal proteinase an interesting subject of investigation in its own right.

A second residue-selective proteinase that may find wider application in sequence analysis is the myxobacter AL-1 protease II studied by Wingard, Matsueda and Wolfe [387]. This enzyme, which cleaves peptide bonds on the *amino* side of lysine residues, has many attributes in common with the staphylococcal proteinase mentioned above. It is extracellular, stable, of low molecular weight (17,000), and exhibits broad specificity with respect to pH. In contrast to the glutamyl-specific enzyme, however, the myxobacter protease does not seem to have an active-site serine, nor does it appear sensitive to sulfhydryl-selective inhibitors. The myxobacter enzyme did not cleave model peptides containing Pro-Lys sequences or C-terminal lysine. A sample of enzyme kindly supplied by Wolfe was evaluated in our laboratory as a means of digesting *Escherichia coli* glutamine synthetase. The hydrolyzed protein was subjected to one cycle of automated Edman degradation and, in addition to the expected amino-terminal PTH-serine [388], nearly theoretical amounts of PTH-lysine were recovered based upon the lysine content of the enzyme. The next stage of the Edman procedure released a wide variety of PTH derivatives, as would be expected. Myxobacter protease II should provide a useful complement to the trypsin digestion at arginines in proteins with blocked lysines mentioned previously. Thus far, a reversible and specific arginine modification as a counterpart to maleylation, and so on, has not proven generally practical.

Finally, it is worth mentioning that several investigators have obtained limited cleavage of proteins at certain arginine bonds by highly purified preparations of thrombin [389, 390]. The specificity of the enzyme resides not only in its preference for cleavage C-terminal to arginine residues, but, in addition, only particular arginine bonds are cleaved. Studies relative to the specificity of thrombin are currently underway in the laboratory of Roger Lundblad, and a recent article by Weinstein and Doolittle [391] deals with this subject. Although as yet no basis for predicting susceptible sites has been established, the list of peptides in Scheme 15 has been kindly submitted to us by Lundblad indicating the

Trp - Ile - **Arg** $\downarrow$ Gly - CMCys

(Ala, Gly, Pro, Ser) **Arg** $\downarrow$ Val-

Arg-Leu-**Arg** $\downarrow$ Asp - Ser

Gly - Gly - Val - **Arg** $\downarrow$ Gly

Phe - Ser - Ala - **Arg** $\downarrow$ Gly

Leu - **Arg**-Pro- **Arg** $\downarrow$ CMCys - **Arg**

Gly - Asn - Tyr - **Arg** $\downarrow$ Asn - Val

Gly - CMCys - Pro - **Arg** $\downarrow$ Gly - Val

Ala - Leu - CMCys - **Arg** $\downarrow$ **Arg** - Ser   Scheme 15

course of thrombic digestion. The phenomenon appears to be all-or-none; an Arg—X bond is either totally cleaved or not at all. Further application of thrombic digestion certainly seems warranted as a means of specific, limited fragmentation of polypeptides.

3.2.2 Selective Chemical Cleavage The new residue-selective proteases mentioned above are a hopeful sign that there may exist a host of such enzymes with differing specificities, waiting to be discovered and applied to the specific generation of large fragments suitable for the automated sequence analysis of proteins. For the most part, however, the increased demand for cleavage procedures that yield large fragments has stimulated more and more interest in the development of specific chemical methods for polypeptide cleavage. An excellent review article on this subject by Spande et al.

[21] presents a comprehensive analysis of this challenging field as it stood in 1970, and is highly recommended in terms of references and the broad scope of the survey to anyone interested in exploring these methods of polypeptide chain scission. For the purposes of this discussion, only a few of the most recent and useful cleavage methods are examined in detail.

3.2.2.1 Cysteine. Without a doubt, the most useful specific chemical cleavage procedure is that described years ago by Gross and Witkop ([392]; see also [21, 393]), which occurs at methionine residues following reaction of a protein or peptide with CNBr. The yields in disruptive acidic media such as 70% formic acid are usually in excess of 90%, especially after pretreatment with a reducing agent to convert reversibly oxidized forms of methionine back to the parent amino acid side chain. The success with the methionine cleavage suggested that similar fragmentations by CNBr might be achieved at S-methylcysteine sites after reduction and methylation of proteins. Interest in a method for cleavage at cysteine residues has been especially intense in the case of structural studies of antibodies, in which the cysteines are positioned in much the same way in molecules with diverse antigen specificity. The job of comparing sequences in a variety of myeloma proteins and antibodies would be facilitated if one could fracture the molecule at these rather invariant sites in the structures. Moreover, in the case of a great many enzymes, a cysteine cleavage method would be a welcome addition to the repertoire of the protein chemist. As pointed out by Spande et al. [21], S-methylcysteine residues (24) do react with CNBr, but by a mechanism different from that observed with methionine. Two possibilities were suggested, as shown in Scheme 16. Low temperatures presumably favor the formation of an oxazolinium bromide 25, which subsequently yields the O-acyl derivative 26, identical to that which would be produced from an N → O shift mechanism. Hydrolysis of the ester bond is complicated by the competing re-formation of the peptide bond through an O → N shift (cf. [394] for references), which would regenerate the original polypeptide backbone with a serine in place of a cysteine. Attempts in our laboratory to produce by this means a papain molecule containing serine in place of the active site Cys-25 have been unsuccessful [395]. The second pathway, indicated by dotted lines in Scheme 16, is favored at higher temperatures and leads to the formation by β-elimination of a dehydroalanine derivative 27. This is hydrolyzed to a fragment with an N-terminal pyruvoyl substituent 28, and one with an amidated C terminus 29.

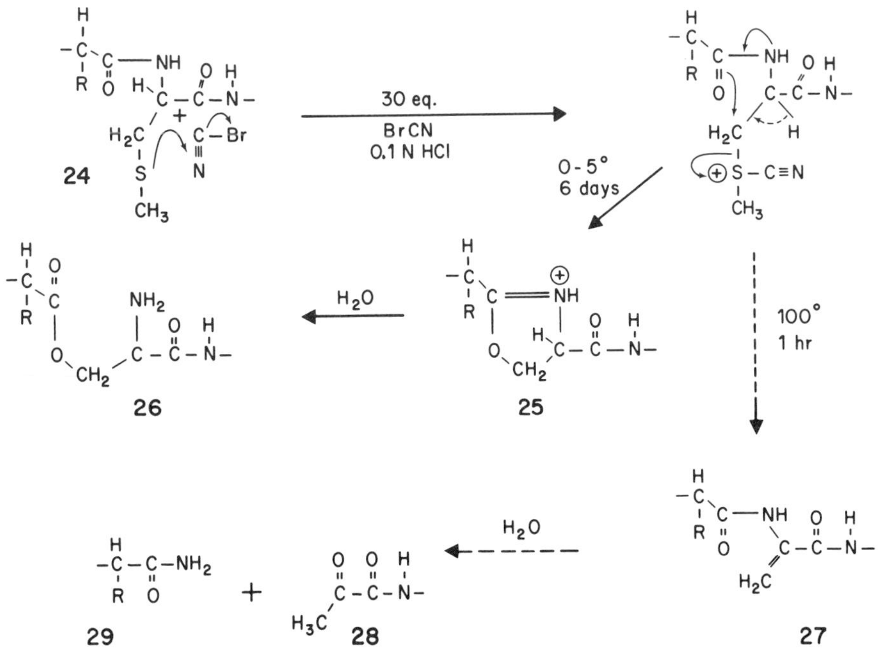

Scheme 16

One of the major drawbacks to further exploration of this selective cleavage procedure derives from the fact that methyl iodide employed in the alkylation step is notoriously nonspecific in its reactions with proteins. This problem has been circumvented by the use of either methyl-*p*-nitrobenzenesulfonate [166] or methyl chloride [289] for the formation of the *S*-methylated proteins. As reported by Awad and Wilcox [396], we observed cleavage in high yield of *S*-methyl glutathione by CNBr at low temperature, with nearly quantitative conversion of *S*-methylcysteine to serine during the course of the reaction. Such was not the case, however, with a model protein, *S*-methyl lysozyme. In this case fragmentation of the molecule was achieved, but conversion of *S*-methylcysteine to serine did not exceed 50% [395]. It appeared therefore that, even at low temperatures, the β-elimination mechanism was a major factor in the cleavage of *S*-methylated proteins by CNBr. The complexity of the mixtures thus obtained argues against this approach to cleavage at cysteine residues.

The importance of a procedure for the quantitative fragmentation of proteins at cysteine sites has sustained continued interest in the matter, and recent developments from Stark's laboratory at Stanford University have all but completely solved the problem. Earlier observations of Catsimpoolas and Wood [397] that CN⁻ causes peptide bond cleavage in cystine-containing proteins led to the conclusion that cleavage was due to formation of S-cyanocysteine which cyclized and cleaved to yield an amino-terminal peptide and an iminothiazolinyl C-terminal fragment. Application of S-cyanylation to the cleavage of polypeptides suffered from the drawback that it was difficult to convert cysteine and cystine residues quantitatively to the S-cyano derivative, and therefore the problem of complex mixtures of cleavage products remained. This disadvantage has been overcome [337, 398] by reacting reduced and denatured proteins with 2-nitro-5-thiocyanatobenzoic acid (NTCB). The synthesis of this reagent recently reported by Degani and Patchornik [292] is shown in Scheme 17. The product, which is formed in almost quantitative

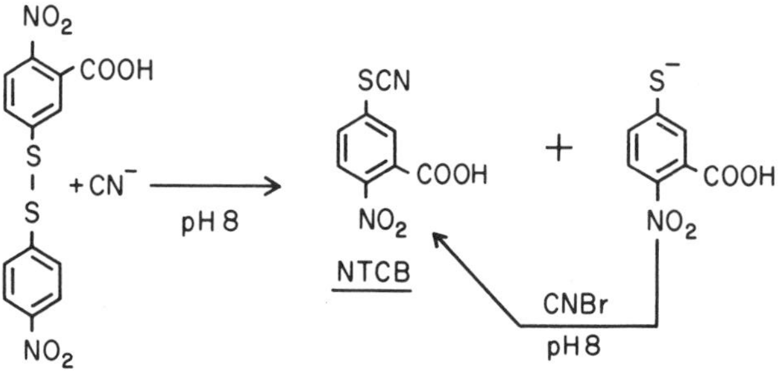

Scheme 17

yield, is quite stable and is easily labeled with ^{14}C by the use of ^{14}CN⁻. Modification of thiols with NTCB is quantitative and selective.

After the problem of quantitative conversion of cysteine and cystine to —SCN was solved, there remained to be discovered conditions that would affort quantitative cleavages at these sites in the protein. A recent article by Jacobson et al. [293] describes the quantitative fragmentation at S-cyanocysteine by exposure to 6 M guanidinium hydrochloride, 0.1 M sodium borate (ph 9.0) at 37°C for 12 hr. This results in the liberation of an amino-terminal fragment and a series of N-blocked iminothiazolidinyl peptides. No side reactions at this slightly alkaline pH were observed. Jacobson et

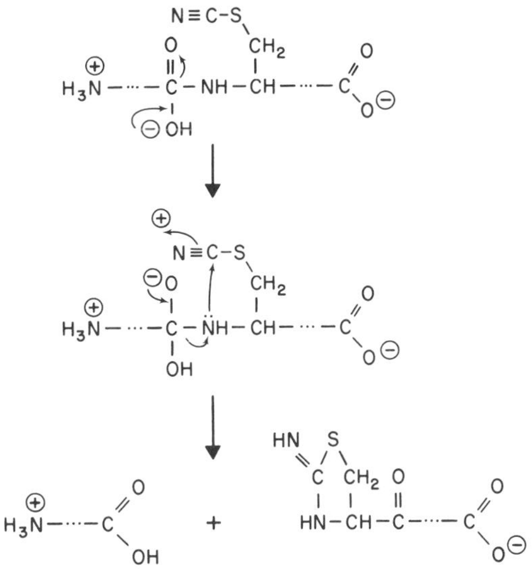

Scheme 18

al. [293] propose the mechanism described in Scheme 18 in which specific hydroxide ion catalysis is followed by concerted peptide bond cleavage and ring closure, without formation of an intermediate acyliminothiazolidine as suggested earlier [21, 397].

The last and perhaps most frustrating obstacle to be surmounted is the question of how to remove quantitatively the blocking N-terminal iminothiazolidinyl substituent or, alternatively, how to open the ring to render the peptides susceptible to Edman degradation. The blocked peptides are resistant both to the Edman procedure and to the enzyme pyrrolidonecarboxylyl peptidase which specifically removes N-terminal pyrrolidone carboxylate residues [399]. As it stands, the procedure will doubtless be of value in placing cysteine residues in proteins and establishing overlaps. Once the problem of the N-terminal blocking group is solved, this cleavage procedure should rank with the CNBr-methionine method both for the quantitative and specific chemical cleavage of polypeptides, and as a means of generating fragments useful for automated Edman degradation.

3.2.2.2 Tryptophan. Chemical cleavages directed toward cysteine and methionine residues are perhaps the most logical choices among the various amino acids that comprise the proteins, owing to the very strong nucleophilic properties of the sulfur atom. Of the remaining side chains, the indole, phenol, and imidazole rings of tryptophan, tyrosine, and histidine, respectively, have received the greatest attention as potential sites for chemical cleavage. The literature on this subject is very extensive indeed, and the interested reader may refer to the review by Spande et al. [21] for a thorough discussion and pertinent references. Oxidizing and brominating agents such as periodate [400] and N-bromosuccinimide (NBS) [401] have been widely employed for the modification of tryptophan, tyrosine, and histidine residues in proteins, as well as for promoting peptide bond cleavage at these sites. The major disadvantage of these cleavage procedures is that they lack specificity, and the yields obtained may be as low as 5%, seldom exceeding 60%. Moreover, undesirable side reactions may be extensive. For example, in addition to its reaction with tyrosine, histidine, and tryptophan, NBS oxidizes cysteine and cystine to cysteic acid and methionine to the sulfone.

Two mild brominating reagents have been described recently which, although still not quantitative in terms of polypeptide cleavage, are nevertheless specific relative to cleavage at tryptophan sites. Omenn, Fontana, and Anfinsen [402] have reported the synthesis and application to tryptophan modification and cleavage of a mild brominating reagent called BNPS-skatole. This stable bromine adduct of 2-(2-nitrophenylsulfenyl)-3-methyl indole is prepared by bromination of the latter compound with NBS and has the following probable structure:

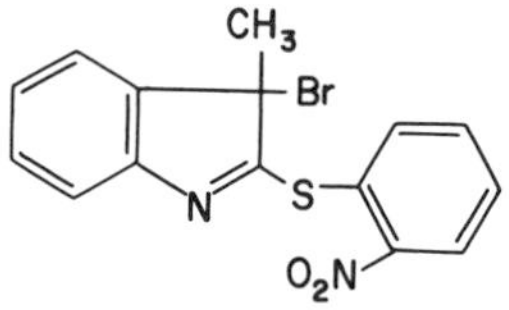

In Table 13 are contrasted the reactions of BNPS-skatole and NBS with a standard mixture of amino acids under varying conditions of time and reagent concentration. These data show that BNPS-skatole can be highly selective for tryptophan relative to the modification of tyrosine and histidine. Cystine was fairly resistant to this reagent, and methionine oxidation proceeded in major yield to the sulfoxide.

**TABLE 13 RECOVERIES OF AMINO ACIDS AFTER INCUBATION
WITH BNPS-SKATOLE OR WITH NBS[a]**

Amino acid	μmoles Recovered		
	BNPS-Skatole		NBS[b]
	10 eq, 30 min	30 eq, 20 hours	20 eq, 30 min
Tryptophan	0.00	0.00	0.00
Lysine	1.01	1.00	1.12
Histidine	1.00	0.99	0.00
NH_3	1.19	1.50	2.55
Arginine	0.98	1.00	0.94
Aspartic acid	1.25[c]	1.40[c]	1.56[c]
Threonine	0.99	1.00	1.03
Serine	0.99	0.98	0.74
Glutamic acid	0.99	0.98	0.89
Proline	0.98	1.00	1.02
Glycine	1.00	1.01	0.98
Alanine	1.00	0.99	1.00
Half-cystine	0.97, 0.86[d]	0.00	0.00
Valine	1.00	1.00	1.00
Isoleucine	0.99	1.00	1.02
Leucine	0.99	1.01	1.00
Tyrosine	1.00, 0.98[d]	0.86	0.00
Phenylalanine	1.01	1.00	0.94

[a] Data from Omenn, Fontana, and Anfinsen [402] with permission. Reaction
mixtures contained 1 μmole of each amino acid in 50% acetic acid.

[b] Degradation of free amino acids by NBS is quantitative at pH 4–5 in 30 min,
but much slower in 50% acetic acid. Under these conditions, mean recovery of
the aliphatic amino acids (side chains not susceptible to attack by NBS) was
80%. For purposes of comparison, recoveries of amino acids are expressed on
the basis of recovery of leucine.

[c] The higher recovery of aspartic acid is caused by coelution of methionine
sulfoxide formed by oxidation of methionine present in the reaction mixture.

[d] 20 eq, 30 min.

Cysteine, if present, is oxidized primarily to cystine, and in low yield
to cysteic acid.

Under more drastic conditions, with large excesses of reagent and
longer reaction times, cleavage at the single tryptophanyl residue of
staphylococcal nuclease proceeded to the extent of about 60%, and
the C-terminal nonapeptide thus produced was isolated in a yield of

15% [402]. Oxidation of about half the tyrosine in the enzyme occurred as well, but no additional cleavages at these sites were observed.

A second reagent for selective fragmentation at tryptophan is 2,4,6-tribromo-4-methylcyclohexadienone (tribromocresol, TBC) recently described in detail by Burstein and Patchornik [403]. A comparison of the reactivities of TBC and NBS with a standard mixture of amino acids is presented in Table 14.

In addition to modification of tryptophan, large molar excesses of TBC lead to the conversion of tyrosine to 3,5-dibromotyrosine, cysteine and cystine to cysteic acid, methionine to the sulfoxide, and

TABLE 14   RECOVERY OF AMINO ACIDS
AFTER INCUBATION WITH TBC OR NBS[a]

Amino Acid	Recovery (%)[b]	
	TBC	NBS
Tryptophan	0	0
Lysine	97	75
Histidine	0	0
Arginine	99	72
Aspartic acid	98	74
Threonine	100	86
Serine	99	58
Glutamic acid	100	71
Proline	99	85
Glycine	100	74
Alanine	98	72
Half-cystine	0	0
Valine	99	80
Methionine	0	0
Isoleucine	99	83
Leucine	100	78
Tyrosine	2	0
Phenylalanine	101	81

[a] Data from Burstein and Patchornik [403] with permission. The reaction mixture contained 0.25 μmole of each amino acid and 15 μmoles of the brominating agent, in 60% acetic acid for 15 min.
[b] As recovered by automatic amino acid analysis.

oxidative losses of histidine. Cleavage of reduced and carboxy-methylated lysozyme occurred at the various tryptophan residues in yields varying from 5 to 60% when the protein was subjected to a 60-fold excess of TBC in 80% acetic acid for 15 min. Under similar conditions no cleavage was observed with ribonuclease, which contains six tyrosines, four histidines, and no tryptophan.

The mechanism by which BNPS-skatole and TBC cleave peptide bonds adjacent to tryptophan is probably the same as or very similar to that of NBS. In the mechanism proposed by Patchornik et al. [401], shown in Scheme 19, three equivalents of oxidative bromine

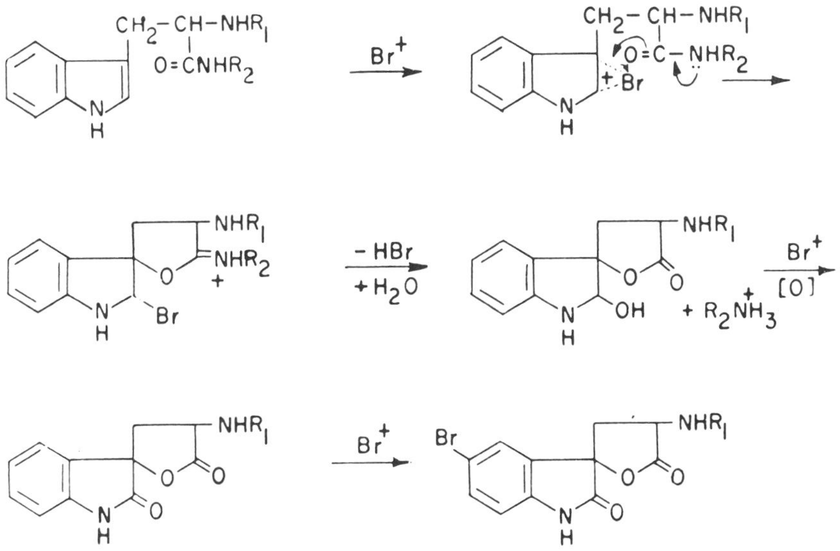

Scheme 19

participate in the cleavage reaction. The first two reactions, which involve bromination and debromination of the indole ring through a series of oxidative and hydrolytic steps, lead to the formation of an oxindole derivative which promotes the cleavage reaction. A stable 5-bromooxindole is produced from the latter by reaction of the third bromine atom. In the case of both TBC [403] and BNPS-skatole, 3 equivalents of reagent are required for the complete modification of a tryptophan. Studies of the reaction with TBC show that only 2 equivalents can be titrated as Br$^-$.

In comparing the two reagents TBC and BNPS-skatole, it appears that the latter may offer certain advantages. Both are employed in high concentrations of acetic acid because of poor solubility in water, are selective for cleavage at tryptophan only, lead to comparable yields of cleavage, and give rise to methionine sulfoxide which may in turn be reduced back to the parent amino acid by thiols. The specificity and mildness of these modifications make them superior to NBS. BNPS-skatole is less destructive toward tyrosine, histidine, and cystine than is TBC, and this may be an important factor in choosing between them for a particular purpose. Moreover, the fact that BNPS-skatole is more stable under a variety of conditions than is TBC argues in its favor. A more pertinent question in regard to structural analysis may be whether or not these methods as they now stand will be of any practical value to the study at hand. The low yields of cleavage are particularly disadvantageous when the protein of interest contains more than four or five tryptophan residues, in which case exceedingly complex mixtures may be encountered. Proteins, such as staphylococcal nuclease, that have a single tryptophan may be ideally suited to such an approach if relatively high yields in fragmentation can be achieved. In any case, progress toward a satisfactory tryptophan cleavage has been promising, and the importance of a quantitative procedure specific for this relatively minor amino acid constituent of proteins should provide impetus for further research in this area.

3.2.2.3 Aspartic Acid and Asparagine. The above discussion centered upon those methods of selective protein cleavage that have been, and promise to be, most useful both in terms of specificity and quantitation of peptide bond rupture. Other approaches have been explored. For example, by maintaining the pH at a level just sufficient to ensure that the β-carboxyl of aspartic acid is in the protonated form, peptide bonds adjacent to this residue are preferentially cleaved in dilute acid [404]. A recent publication of Bornstein and Balian [405] describes conditions for cleavage of Asn—Gly bonds by hydroxylamine. Reduced and carboxymethylated ribonuclease subjected for 2 hr at 45°C to 2 M hydroxylamine, 0.2 M K_2CO_3 (pH 9.0), underwent peptide scission between Asn-67 and Gly-68 (about 60%), and to a lesser extent between Asn-34 and Leu-35. These investigators proposed that cyclization of asparaginyl side chains may lead to intrachain cyclic imides susceptible to nucleophilic attack and cleavage by hydroxylamine. At pH values higher than 9, competitive nucleophilic attack by OH^- leads to saponification of the cyclic imides, and therefore lower yields of the cleaved products. This probably accounts for the fact that uncleaved protein isolated after hydroxylamine treatment is resistant to further cleavage by the reagent, and also the observation that protein subjected to pH 11 prior to hydroxylamine is not cleaved.

Although the yields obtained by this cleavage are not as high as one would hope, and the specificity for Asn–Gly bonds is not absolute, the low frequency of these linkages in proteins may justify attempts by this means to provide large fragments amenable to automated Edman degradation. Titani et al. [406] have reported success with this procedure in the fragmentation of thermolysin, but our attempts to cleave bovine liver rhodanese between a known Asn–Gly bond have not yielded the expected, or indeed, any products [407].

One of the peptide bonds involving aspartate, namely, Asp–Pro, appears to be especially labile to acidic conditions. During the course of their work on the covalent structure of glutamate dehydrogenase, Piskiewicz, Landon, and Smith [408] observed partial hydrolysis of an Asp–Pro bond which accompanied cleavage of the enzyme by CNBr in 70% formic acid. We have encountered two such acid labile Asp–Pro bonds in yeast inorganic pyrophosphatase [409, 410]. Minor cleavages (about 15-20%) occurred at these two sites during treatment with CNBr in 70% formic acid, and the purified products generated thereby were extremely valuable in terms of the information obtained from them by automated Edman degradation. Therefore a side reaction that might have been an annoyance proved in fact to be fortuitous and enabled us to extend our knowledge of the enzyme sequence with little extra effort. In view of the low frequency of Asp–Pro linkages in proteins, and the fact that cleavages at these sites can provide fragments amenable to automated Edman procedures, further exploration of means to increase the yields of cleavage seems highly warranted.

3.3 Sequential Degradation of Proteins

This final section deals with current approaches, both by chemical and enzymic means, to the sequential removal of amino acids from proteins and peptides. Previous reviews and texts related to this subject [16-20, 411, 412] provide ample documentation for the reader who is interested in detailed experimental procedures for many of the methods discussed. To date, chemical methods for establishing C-terminal sequences have not proven to be very satisfactory and, since the subject has received comprehensive treatment in the recent reviews by Stark [20, 413], no further discussion is presented here. The application of mass spectroscopic analysis to the determination of covalent structure in peptides has met with some success, especially in the analysis of complex structures and of peptides available only in submilligram quantities. Although this method has proven most valuable with peptides of 10 residues or less, a recent report by Nau, Kelley, and Biemann [414] describes the complete sequence elucidation by mass spectrometry of a 20-residue CNBr fragment. This approach, still in its infancy, has been treated in recent articles by Forster et al. [415] and by Leclercq et al. [416]; the interested reader is referred to these publications for further references pertinent to this subject.

3.3.1 **Chemical procedures from the Amino Terminus** Examination of the current volume of *The Atlas of Protein Sequence and Structure* [15] will reveal that in the last 15 years an extensive body of information has been derived relative to the primary structures of a host of proteins and peptides. It is safe to say that the vast majority of these sequence data was obtained by the application of a stepwise chemical degradative scheme devised in the early 1950s by Edman [417]. Although other methods for sequential analysis from the amino terminus have been devised (e.g., [418, 419]), that of Edman has withstood the test of time, and his procedure, together with some useful modifications based upon the same general idea, are in widespread use throughout the world. The degradative reactions shown in Scheme 20 may be considered in three stages: (1) coupling of the protein or peptide with an isothiocyanate, (2) cyclization and cleavage of the first amino acid as a thiazolinone, and (3) conversion of the latter to a thiohydantoin. In the standard Edman procedure, R is a benzene ring, and the coupling reagent is phenylisothiocyanate (PITC). The first reaction at pH 9–9.5 yields a phenylthiocarbamyl (PTC) peptide which is then cleaved to give an anilinothiazolinone (ATZ) which is finally converted in the third stage to the PTH amino acid.

Edman's innovation was based upon a similar series of reactions, first devised by Bergman and Miekeley [420] and developed further by Abderhaden and Brockmann [421], in which phenylisocyanate was employed as the coupling reagent. The vigorous acidic conditions required for cyclization and cleavage of the phenylcarbamyl peptide, however, led to hydrolysis of peptide bonds, and the method was not really of practical value until Edman replaced phenylisocyanate with the sulfur derivative (PITC). The PTC peptides undergo cyclization

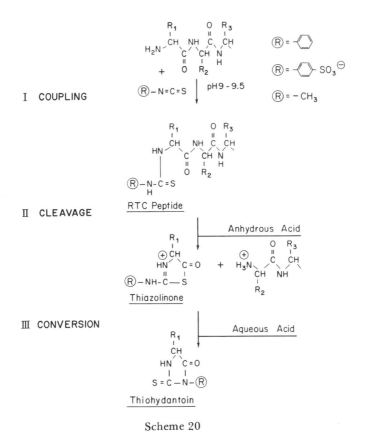

I   COUPLING

II CLEAVAGE

III CONVERSION

Scheme 20

and cleavage in anhydrous acid under mild conditions in which peptide bond scission is minimized. Several articles dealing with this subject have been published by Edman and co-workers [422–425], and the method is quite simple and straightforward. The coupling reaction is performed at a pH high enough to favor existence of the reactive unprotonated form of the amino group and yet not exceeding 10, at which pH the PITC is unstable. A wide variety of volatile buffer solvents has been employed that favor solvation of both the peptide and PITC. Among these, 50% aqueous pyridine and dimethylallylamine (DMAA) are in widespread use. Coupling is usually complete in 30 min at 50°C. After removal of reagent and reaction side products by extraction with benzene, the dried residual peptide is exposed to an anhydrous acid such as trifluoroacetic acid for 5-10 min at 45-50°C. The thiazolinone thus produced is removed by extraction (ether or butyl chloride), and the dried peptide is ready for another cycle of the procedure. The ATZ is converted to the PTH amino acid by exposure for 10 min at 80°C to 1 N HCl, and the latter is identified by any of a number of methods described later in this section.

The brief outline presented above touches upon only some of the more important features of the general degradative scheme; pertinent developments in this regard are mentioned in the ensuing discussion. Most investigators have made slight modifications in their own personal adaptations of the method (e.g., [426]), and the search for "superadditives" to give prolonged mileage into a peptide chain continues. One important factor, recognized by Edman and several other investigators who sought to make the procedure more efficient, was the necessity for the strict exclusion of oxygen during the degradative cycles. In the presence of oxygen, PITC and PTC peptides undergo oxidative desulfuration resulting in the formation of isocyanate derivatives. These are effectively blocked from further degradation, since they are not cleaved under the mild conditions sufficient for cleavage of PTC peptides. For years, investigators who performed the Edman degradation manually have done so by conducting all stages of the procedure in an atmosphere of nitrogen or under vacuum. Clearly, the complete elimination of oxygen is not possible under almost any conditions, but is especially inefficient with manual methods. Although manual sequence analysis of more than 20 residues has been reported [cf. [426]), degradation of a peptide is usually limited to 8 to 15 cycles and the manual procedures have been most effective with smaller fragments.

The shortcomings of the manual method, primarily with respect to oxidative losses in yield at each step, prompted attempts to automate the procedure and thereby render it more efficient. In 1967, Edman and Begg [354] reported details for the construction and application of a protein "sequenator" by means of which they were able to confirm the structure of sperm whale myoglobin through the first 60 residues. In order to obtain information as extensive as this, it is necessary that the repetitive yield at each step exceed 98%. The fulfillment of this requirement depended upon rigorous purification of all solvents and reagents and the most meticulous care to exclude oxygen from the entire system. At first sight, automation of the Edman procedure might not appear to pose any particularly overwhelming problems. However, in view of the stipulations mentioned above relative to exclusion of oxygen, and the fact that some of the reagents and solvents are extremely corrosive in addition to being highly flammable, technical problems relating to the construction of the reaction vessel, valves, seals, tubing, and so on, represented a considerable obstacle to automation. The reaction vessel itself is a spinning glass cup into which solvents and reagents are added to a film of the dried protein or peptide in very precisely controlled quantities. It is vital that the surface area of the cup covered by the reaction mixture be constant from cycle to cycle at any particular step of the degradation. Precise control of the cup rotational speed is therefore also of importance. The cup is enclosed in a system that can be evacuated to less than 50 μ of Hg, and is also fitted with a nitrogen inlet tube for drying of the sample film at various stages. The temperature of the reaction mixture is maintained at 50–55°C during each cycle. Removal of reaction by-products and excess reagents is effected by passing an appropriate solvent over the dried sample film for various lengths of time. The solvents rise up the wall of the cup and into a groove at the top into which is fitted a piece of tubing for conveying the effluent to waste or to a refrigerated fraction collector. In this way, the ATZs extracted in butyl chloride are collected sequentially in a volume of about 3 ml, and these are dried under a nitrogen stream and *in vacuo* during the next cycle. In order to avoid precession of the cup, which would lead to turbulence in and bumping of the reaction mixture, this component must be very carefully machined. It is most likely this consideration that has discouraged many investigators from attempts at building their own "sequenators," although Waterfield, Corbett, and Haber [427] have reported details for the construction of a protein sequencer designed along these lines. In considering the various instruments currently available on the market, the Beckman 890-C Protein-Peptide Sequencer probably ranks at the top in terms of quality of the instrument and continuing research and development expended in its improvement.

In addition to the earlier article by Niall [428], Fietzek and Kuhn [411] have recently published a very thorough review comparing the Beckman sequencer to the sequenator of Edman and Begg [354], in which various aspects of programming and reagent choices are discussed. Putnam [412] has also written an article pertinent to this subject. Our account is therefore justifiably limited; the devotion of as much time to the subject as has been expended seems appropriate in view of the ever-increasing application of spinning-cup protein sequencers to the determination of primary structure. These sequencers are not the only means developed for automating the

Edman procedure. As mentioned previously, spinning-cup instrumentation was originally intended for large protein fragments with which extractive losses during washing steps with organic solvents would be minimized. At about the same time as commercial sequencers began to appear on the market, Laursen [429, 430] reported the development of a solid-phase method for performing the Edman degradation, by means of which peptides could be sequenced with great efficiency. Whereas in conventional spinning-cup sequencers the sample is immobilized as a thin film, the Laursen procedure involves covalent attachment of the protein or peptide to an insoluble matrix or resin. After reversibly blocking amino groups with t-butyloxy-carbonyl azide, carboxyl groups activated by carbonyldiimidazole are coupled to an aminopolystyrene resin. The Edman procedure is then performed automatically by pumping reagents and solvents through a column of the peptide-resin. A complete description of the instrumentation and other particulars relevant to the application of the solid-phase method may be found in Vol. XXV of *Methods in Enzymology* [431].

There are several advantages and disadvantages associated with the solid-phase approach. The sequencing apparatus developed by Laursen is simple in design and operation and may be constructed from commercially available components at a fraction of the cost of spinning-cup machines. The reduction in moving parts and the versatility of the apparatus are factors definitely in its favor. Furthermore, the strict requirement for reagent and solvent purity essential for the high repetitive yields obtained in spinning-cup sequencers is less important in the solid-phase method. This amounts to a considerable reduction in operating costs on an annual basis.

As discussed by Laursen [431], the shortcomings of the solid-phase method relate primarily to the mode of coupling of peptide to the insoluble support. Although the yields in this step are often good (70–100%), it is not possible to avoid attachment at side chain, as well as C-terminal carboxyl groups. With glutamic acid one merely observes a gap in the sequence, since the cleaved glutamyl ATZ remains attached to the resin. In the case of aspartic acid, the degradation is stopped completely. The mode of attachment of peptide to resin has been, and continues to be, an area of intense investigation. With tryptic peptides it is possible to convert all carboxyl groups to amides and then regenerate the C-terminal carboxyl by cleaving again with trypsin [431]. We have attempted in our laboratory to completely amidate the carboxyls of a protein prior to tryptic digestion but, owing to poor solubility, the resulting protein derivatives are not generally amenable to proteolysis. Tryptic peptides terminating in lysine have been coupled via the ε-amino to resins, with the advantage that prior chemical modification is not required [431]. Other problems with the Laursen method relate to difficulties in the identification of serine and threonine [431]. All things considered, the raison d'être of the solid-phase approach, aside from its simplicity and economy, resides in its application to peptide sequencing. For reasons discussed above, the method would not be applied conveniently as a general approach to the sequence analysis of proteins.

In comparing and contrasting these two modes of automated Edman degradation, it appears that, in terms of more general applicability to samples wide-ranging in molecular weight, the spinning-cup approach is the method of choice. One of the most attractive features of the latter is that no chemical modification of the sample is required prior to analysis although, as discussed later, current approaches to small-peptide sequencing in these systems do rely upon derivatization procedures. Since it is possible to add a precisely known amount of sample to the cup, quantitation of the degradation is facilitated. Because no coupling of sample to an insoluble support is required, the method works well with large polypeptides ranging in size from 50 to 500 residues, for which extractive losses are minimal. One may hope to succeed in identifying 30 to 50 residues during the analysis of such proteins, and the strategy has been to locate residues at which specific cleavage procedures may be applied to generate large fragments for further sequence studies. Thus far application of the sequencer to the determination of the complete sequence of a protein has had its greatest utility in studies in which the strategy was to isolate a few large fragments. In practice, however, if a complete structure is to be elucidated, one is usually faced ultimately with the problem of isolating and sequencing peptides 30 residues or less in length. With samples of this size, extractive losses due to solubility of the peptides in the organic solvents may be prohibitive to extensive structural analysis in the spinning-cup sequencers even with the application of special peptide programs with volatile buffers [411]. Therefore the chemical modification of peptides to render them more polar, hence more resistant to extraction, has been a subject of great interest. Succinylation of amino groups has been explored by Niall et al. [432] as one approach in this direction, but ideally one would wish to introduce a group of even lower pK. Braunitzer et al. [433] have recently proposed the use of sulfonated PITC (SPITC) as the coupling reagent in the first step of Edman degradation (Scheme 20,

R = —⟨ ⟩—SO_3^-). All the ϵ-amino groups, as well as the α-amino

groups, would be modified by the reagent to yield residues with strongly acidic side chains. Cleavage and removal of the first residue as the sulfonated ATZ are performed manually, and then the automated degradation is continued with PITC. The first residue may be identified by acid hydrolytic back-conversion as described later in this section. This method obviously depends on the presence of lysine residues, or amino groups, close to the carboxyl terminus of the peptide in question, so that the sample remains insoluble in the organic extractants as long as possible during the course of the degradation. Conversion of sulfhydryl groups to S-aminoethyl derivatives would be especially useful in this regard, because of the introduction of additional sites for substitution by SPITC. In our experience, use of the Braunitzer reagent has enabled us to proceed to the end of a lysine-terminal tryptic peptide containing 25 residues utilizing the spinning-cup sequencer with a DMAA-peptide program. In similar attempts with the unmodified peptide, meaningful interpretation was impossible after 13 cycles owing to losses of peptide from the cup.

Since not all peptides contain side-chain amino groups, it would be desirable to have in the repertoire a means of rendering peptides polar that would be more generally applicable. An attempt in this direction has been reported recently by Foster et al. [434], in which the carboxyl groups of nonpolar peptides are coupled to the amino group of 2-amino-1,5-naphthalene disulfonic acid (ANS) in the presence of a water-soluble carbodiimide. Since the pK of the amino group in ANS is less than 2.5, the coupling reaction may be performed at pH 4, under which conditions the α-amino group of the peptide is unreactive and does not require protection. In the spinning-cup degradation of native and modified Gly-Val-Pro-Gly, the percent recoveries of PTH derivatives after four cycles of the Edman procedure were, respectively, 80, 5, 0, and 0 for the unreacted peptide, and 90, 44, 16, and 8 for the more highly charged derivative [434]. The peptides studied by these investigators did not contain β- or γ-carboxyl groups, which would also be expected to couple with ANS. Presumably, a higher incidence of substitution would favor the polar nature and film-forming properties of the peptides, but this is an area yet to be evaluated.

Having discussed certain broad aspects of the instrumentation involved in automating the coupling and cleavage phases of Edman degradation, and before mentioning specific applications of the sequencer, attention is focused briefly on contemporary procedures for the identification and quantitation of PTH amino acids as they are sequentially removed from the peptide chain. Scheme 21 is a

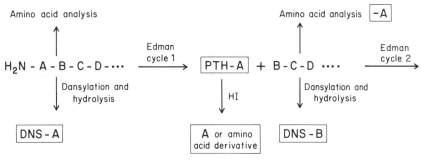

Scheme 21

schematic diagram representing the various modes of monitoring a sequence analysis utilizing the Edman degradation. Two methods, the subtractive Edman [435] and the dansyl (DNS) Edman [436] have been and continue to be of value in manual sequence applications but, because they involve manipulations with the residual peptide, are not generally used in automated sequence analysis. In these procedures, an aliquot of peptide is removed after each stage of the degradation for amino acid analysis or dansylation. One then follows the sequence analysis by the identification of the new N-terminal residue as its DNS derivative, or by the loss of an amino acid after each cycle. Although the subtractive method is quantitative, it is useful only with fragments of relatively simple amino acid composition. The dansyl method is not quantitative, but is quite sensitive; identifications are made by thin-layer [437] or paper electrophoretic [438] procedures.

With the automated Edman approach, the sequence analysis is based upon the chromatographic identification and quantitation of the PTH amino acid (direct Edman) or of the parent amino acid derived therefrom by hydrolytic back-conversion. Thin-layer chromatography has been employed widely as a means of identifying most of the PTH amino acids [439]. Laursen [440] has rendered this method quantitative by employing radioactive PITC in the coupling step. The labeled PTH amino acids recovered at each step are then diluted with a standard mixture of nonlabeled derivatives, thin-layer chromatography is performed, and all the spots are counted in order to detect the radioactive residue from the peptide. Each laboratory involved in sequence analysis utilizes methods that give a high level of confidence in the final assignments. We rely upon thin-layer methods for corroboration of results obtained by two quantitative procedures, gas chromatography and hydrolytic back-conversion. With this battery of methods, we feel that this confidence is merited. There are in use currently a number of programs for the gas chromatographic analysis of PTH amino acids. Pisano and Bronzert [441] reported conditions for separation of PTH amino acids and their silylated derivatives which, with or without slight modification, are widely employed for this purpose. In conjunction with their development of the sequencer, Beckman Instruments has devoted considerable attention to the gas chromatographic analysis of the PTH derivatives and several useful guidelines are published in their bulletins "Fractions" and "In Sequence." In Figure 4 is shown a series of gas chromatograms obtained during the course of automated Edman degradation of yeast inorganic pyrophosphatase [70]. The sequence analysis has now been extended to 36 residues, and the chromatograms serve to illustrate the ease of quantitation of residues far into the sequence. The gas chromatographic method works well with most of the PTH derivatives; those of glutamic and aspartic acids must be silylated, and this modification permits distinction of isoleucine from leucine. Moreover, silylation improves yields of serine, threonine, glutamine, asparagine, and tyrosine. Basic amino acids and derivatives of cysteine are perhaps those most difficult, if not impossible, to quantitate by gas chromatographic procedures. Identification of PTH-arginine and PTH-histidine has relied upon spot tests [442], although the latter residue and PTH-lysine may be hinted at by gas chromatography. PTH-cysteinyl derivatives pose a special problem, and various approaches have been used to identify this residue in a sequence; quantitation is more difficult. Carboxymethylcysteine gives a PTH derivative that is very poorly observed on gas chromatography, although we have had success with the amidated derivative during the sequence analysis of a structural protein, choricin, from the eggshell of the North American giant silk moth (*Anthereae polyphemus*). Other PTH-cysteinyl derivatives that have been useful in terms of the gas chromatographic approach are

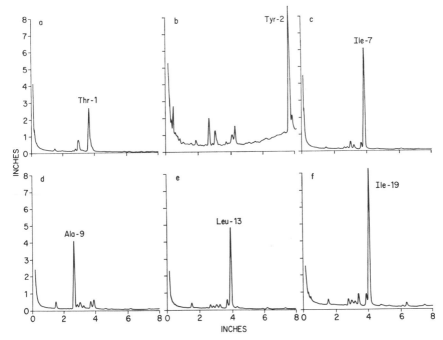

Figure 4   Gas chromatographic patterns obtained from several PTHs produced during automated Edman degradation of alkylated yeast pyrophosphatase. Graphs are exact reproductions of the experimental data plotted as inches of pen deflection versus inches of peak elution position at a fixed chart speed with reference to the point of injection. All samples contained 8% of the total PTH removed at a given cycle, and all the data were obtained at an attenuation of 1600, with the exception of Ile-19 (*f*) which was run at an attenuation of 800. Tyr-2 (*b*) was determined as the trimethylsilyl derivative. From Heinrikson et al. [70] by permission.

S-methyl- [166] and *S*-pyridylethyl- [290, 291] substituted amino acids. Some workers convert cysteines to ^{14}C-labeled derivatives prior to sequence analysis and rely upon radioactivity measurements for identification of the cysteine positions. We have had limited success with this approach, insofar as the radioactivity seems to appear to some extent at every cycle.

With the exceptions noted above, gas chromatography can serve as a rapid means for identifying and quantitating PTH amino acids as they are sequentially released from a polypeptide. We routinely employ this procedure to follow the degradation of a protein; the "sequencer" yields ATZs at the rate of about one per hour, and at this rate one can keep pace by use of gas chromatographic identifications. However, we invariably encounter "gaps" in the sequence, or positions that are questionable, and to resolve these questions, we hydrolyze the appropriate PTH derivatives with HI to convert them back to the parent amino acid or some identifiable derivative. This valuable method, developed and refined by Smithies and co-workers [443], provides in turn a second independent method for obtaining quantitative as well as qualitative information regarding the PTH amino acid in question. Amidated acids of course are hydrolyzed to glutamate and aspartate, and isoleucine is quantitated by the sum of isoleucine plus alloisoleucine. This method allows the precise quantitation of histidine, lysine, and arginine, and α-aminobutyric acid and alanine are obtained in high yields from threonine and serine, respectively. S-Alkyl cysteine derivatives are also converted to alanine, and distinctions must be based upon other methods as discussed above; we have obtained good yields of cysteic acid following hydrolysis of the PTH amino acid. The application of one or both of these quantitative approaches yields reliable data for the sequential identification of the PTH amino acids during the course of sequence analysis as long as the background "noise level" is not prohibitive.

From the foregoing discussion, it is clear that direct identification of the amidated residues, glutamine and asparagine, can be made only by analysis of the PTH derivative itself. The subtractive and DNS Edman procedures and the back-conversion method of Smithies et al. [443] result in deamidation, and the presence of amidated residues must be inferred indirectly, for example, by the electrophoretic behavior of the fragment in question. Although this is practical with small peptides where few residues of this kind must be distinguished, it is obviously impossible with large fragments and proteins that are subjected to automated degradation. Identification of glutamine and asparagine residues in the latter is best made by thin-layer or gas chromatographic analysis of the PTHs. Since these derivatives are often difficult to identify as such, tentative assignments may often be confirmed by back-conversion in HI. If good yields of the corresponding acids are thus obtained, and if these were not present prior to back-conversion, it is highly likely that the original PTH derivative was amidated.

Another approach to this problem, described recently by Gibson and Anderson [444], involves amidation of all free carboxyl groups in the sample by coupling with glycine ethyl ester in the presence of a water-soluble carbodiimide. Aspartic and glutamic acids are thereby converted to β- and γ-substituted derivatives:

$$\begin{matrix} O & H & & O \\ \parallel & \mid & & \parallel \\ -C & -N-CH_2- & C & -O-CH_2-CH_3 \end{matrix}$$

β or γ

Hydrolytic back-conversion of the PTH derivatives afforded by automated Edman degradation of the derivatized proteins or peptides yields either glutamic or aspartic acids and an equimolar quantity of glycine. Amidated residues would give only the corresponding acid. This method has been successfully employed in distinguishing between glutamic and aspartic acids and their respective amides in ribonuclease, but its general applicability to other proteins has yet to be assessed.

Before closing this section, it is worthwhile to consider briefly how automated Edman degradation is being applied at present to a variety of structural problems, and where further development in this field is likely to take us. The main thrust of Section 3.2 was directed toward methods that yield large fragments, because, to date, spinning-cup sequencers have found their greatest utility with peptides greater than 20 to 25 residues. Attempts, mentioned earlier, to make the procedure generally useful for small peptides have been valuable and hold promise for eventual success, but currently the vast majority of sequence information obtained in this fashion has come from analysis of proteins and large peptides derived therefrom. In his recent review, Putnam [412] outlines approaches and strategies for the use of sequencers, with emphasis on application to the structural analysis of immunoglobulins.

One of the major areas of productivity with respect to automated degradative procedures has been in comparative studies of amino-terminal segments of homologous proteins. Comparative sequence information of this kind derived from analysis of hemoglobins, cytochromes, and families of proteolytic enzymes has had an immense impact upon our ideas of evolutionary processes. This very important field of study has been the subject of numerous reviews (cf. [445, 446]) and is not discussed further here.

Another important use of the sequencer is as an exacting criterion of protein size and homogeneity. We have employed automated Edman degradative procedures in studies of the subunit structures of yeast inorganic pyrophosphatase [70] and E. coli glutamine synthetase [388]. With the latter enzyme, it was proposed (cf. [447]) that the native structure comprised 12 identical subunits of molecular weight 50,000. Strong evidence in support of this contention was obtained by structural studies which revealed a single N-terminal sequence of 20 residues and a single C-terminal sequence [388]. Yields of the PTH amino acids were in accord with a molecular weight of 50,000. In the case of the pyrophosphatase, there was considerable disagreement as to the size and subunit composition of the enzyme, partly as a result of the fact that some studies were carried out with partially cleaved preparations (cf. [70] for pertinent references). Our contention, based upon molecular weight and peptide mapping data, was that the native enzyme is composed of two identical subunits approximately 30,000 in molecular weight. This was borne out by sequence analysis that gave a single N-terminal sequence for the first 20 residues with yields in accord with what would be expected in a protein of this size. These findings served as a basis for the complete structural analysis of the pyrophosphatase currently in progress; information subsequently derived is in agreement with the proposed two identical subunits [409].

Finally, the sequencer has played a central role in the complete covalent structural analysis of several proteins. Putnam [412] gives an informative table summarizing strategies used for several protein hormones of less than 100 residues. In bovine neurophysin II, 80 of the 97 residues were determined by automated degradation [448]; 66 of the 84 residues were similarly identified in bovine parathyroid hormone [449]. The latter result was confirmed by independent means [432]. In these examples, the sequences were completed by manual analysis of smaller peptide fragments, but the brunt of the attack was borne by the sequencer. Brandt and Von Holt [450] have reported the complete sequence of 136 residues in histone F3 from chicken erythrocytes obtained solely by the amino- and carboxyl-terminal sequence analysis of the protein and large fragments derived therefrom by limited specific-cleavage procedures. An example of how the automated approach can serve to corroborate and facilitate conventional procedures of structural analysis is provided by the work of Titani et al. [406]. These workers isolated and subjected to automated sequence analysis six fragments from thermolysin, and thus placed 189 of the 316 residues in the molecule. In so doing, they confirmed large portions of the sequence determined by conventional means, provided evidence for the assignment of certain overlaps, and gave direct evidence for identification of glutamine, asparagine, and tryptophan residues.

An interesting article in addition to that of Putnam [412] on the strategy of automated sequence analysis is one recently published by

Hermodson et al. [291]. The Seattle workers in the laboratory of Hans Neurath have provided many new insights regarding the effective use of automated sequence methods, and efforts are being expended in other laboratories to further automate the procedure. One can certainly envision extension of the degree of automation to include conversion of the ATZs and subsequent analysis of the PTH derivatives. This would be of extreme importance with respect to the technical manipulations currently demanded by the method. Ideas have been put forth regarding coordination of the sequencer and a mass spectrograph for analysis of the amino acid derivatives. In this application the use of methylisothiocyanate (Scheme 20, R = $-CH_3$) rather than PITC has been proposed, since the resulting derivatives give good molecular ions and simple mass spectra [451]. It seems likely that further developments in the near future will provide instrumentation such that one need only apply a sample to the spinning cup, set the appropriate dials, and then return in 24 hr to obtain the sequence on a computer display giving amino acid and yield at each step of the degradation. In the meantime, there is every indication that with further research and refinement in this area, especially in regard to small-peptide sequencing, Edman-Begg type sequencers will be centrally involved in future primary structure analysis of proteins and peptides wide-ranging in molecular weight.

3.3.2 **Enzymic Procedures for Sequence Analysis**   In recent years several enzymes have been described which are of particular interest to the field of protein sequence analysis. The determination of amino acid sequences based upon the stepwise enzymic liberation of amino acids or dipeptides from one end of the polypeptide chain is subject to difficulties in interpretation which arise in large measure from the kinetic complexities of the hydrolysis. Carboxypeptidase A, for example, exhibits strong specificity toward particular C-terminal residues [452], and meaningful interpretation of results obtained by digestion with this enzyme seldom extends beyond five residues. Furthermore, it is essential that the exopeptidase be free of any protease activity that would give rise to new terminal sites of attack and thus further confound evaluation of the results. Despite these reservations, the use of enzymes that sequentially degrade proteins from one or the other end of the chain has been of immense importance to structural studies and continues to be the most widely employed means of establishing C-terminal sequences to complement results of Edman degradation and to establish overlaps of peptide fragments.

Before mentioning a few of the most recently described enzymes of importance to sequence analysis, note should be made of an enzyme that removes pyrrolidonecarboxylic acid (PCA) from the N-terminus of peptide chains. The existence of this residue in a larger number of proteins constitutes a major obstacle to sequence analysis, since the blocked chains are resistant to Edman degradation. Moreover, many peptides generated by proteolytic means in the course of structural analysis may contain N-terminal glutamine and subsequently undergo spontaneous cyclization to PCA-blocked fragments. We have observed this with a tryptic cysteinyl peptide from glutamine synthetase, in which cyclization of N-terminal glutamine occurred over a 6-week period at $-4°C$ [453]. In a current review of the subject, Doolittle [399] presents details for the isolation and use of an enzyme, pyrrolidonecarboxylyl peptidase, that specifically removes PCA from the N terminus, thereby generating a fragment upon which further structural studies may be performed. Although information with regard to the enzyme is still scant, it appears to work best on small fragments. However, Edelstein, Lim, and Scanu [454] have had success in removal of PCA from the 78-residue *Macacus rhesus* apoprotein from the high-density lipoprotein, and the residual fragment thus produced was amenable to Edman degradative procedures. No doubt the pyrrolidonecarboxylyl peptidase will find increasing application to this particular problem of structural analysis.

While attention is focused on the N terminus, brief mention might be made of the aminopeptidases (cf. [455]). Use of these enzymes as tools for sequence determination has been overshadowed by and large by the much more rigorous Edman method. However, aminopeptidase M [456] has been of particular value for the complete hydrolysis of small peptides under conditions in which sensitive amide linkages and tryptophan residues are not destroyed.

Of greater interest and potential with respect to sequence analysis are the dipeptidyl aminopeptidases (DAP) which have been characterized extensively in the laboratories of Callahan [457–459] and Fruton [460–462]. These enzymes catalyze the sequential removal from the N terminus of a series of dipeptides. Four such enzymes with distinct substrate specificities have been studied to date [457], but the one designated DAP-I has the broadest specificity, hence is of greatest utility as a sequencing tool. The recent review by McDonald et al. [459] describes methods for the preparation of DAP-I from beef spleen, and the application of this enzyme to the sequence analysis of several polypeptides. Scheme 22, taken from that article, indicates some aspects of the specificity of hydrolysis. The enzyme can remove dipeptides having penultimate residues with polar or nonpolar side chains; basic residues are particularly favorable sites of attack. The nature of the N-terminal residue appears to have greatest influence on rates of hydrolysis, especially arginine and lysine which terminated cleavage of β-corticotropin, glucagon, and secretin. Sequences of acidic residues retard hydrolysis, and the enzyme does not appear to cleave peptide bonds involving proline. Tripeptides may or may not be susceptible to cleavage by DAP-I.

As mentioned by Callahan et al. [463], use of DAP-I for sequence work has certain attractive features, especially in that after a series of dipeptides has been isolated and ordered, the original fragment can be shortened or lengthened by one residue and hydrolyzed again to yield a second set of overlapping dipeptides. The sequence can thus be established from a knowledge of the compositions alone. The major obstacles to the procedure relate to isolation of dipeptide fragments and their correct ordering in sequence. Separation procedures involve use of an amino acid analyzer and paper chromatography (cf. [463]), and standard systems have been developed by means of which, in many cases, the peptide can be identified on the basis of chromatographic behavior alone. Ordering of dipeptides is unnecessary if the overlapping set of fragments is produced unless, as is often the case, there are duplicate amino acids present. Peptide impurity, even at low levels, may lead to erroneous sequence analyses. In view of the several complexities involved with this procedure, it remains to be seen whether it will ever prove to be a viable alternative to the highly successful chemical methods for deriving N-terminal sequences. It may be that application of mass spectrometric procedures to the identification of the sequentially released dipeptides can increase the potential of this approach to sequence analysis.

The elucidation of C-terminal sequences in proteins and peptides by means of chemical procedures analogous to those successfully employed at the N terminus has been a subject of intense inquiry (cf. [20, 413]). Thus far, however, the most fruitful approach to C-terminal studies has been through analysis of the kinetics and stoichiometry of release of amino acids during digestion with carboxypeptidases (CPase). The literature contains manifold references to the application of mammalian CPases A and B for this purpose, and these enzymes have been exhaustively reviewed in

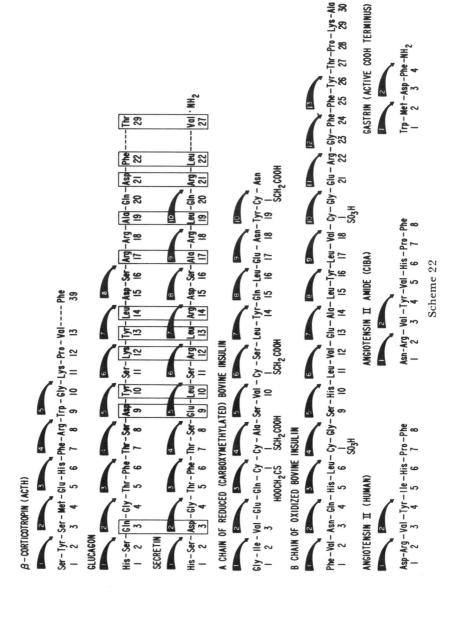

Scheme 22

terms of specificity and methodology [452, 464]. The two enzymes complement one another insofar as CPase B is relatively specific for removal of arginine and lysine and, with the exception of proline, the A enzyme removes most other amino acids, albeit at rates that may vary considerably. The pH optimum for these exopeptidases lies between pH 8 and 9, although CPase A has been used at pH 5–6 for more efficient removal of glutamate and aspartate [465, 466]. The latter enzyme has been particularly useful in the C-terminal analysis of CNBr fragments; at pH 8.5 there is an equilibrium between homoserine and its lactone, and the liberation of homoserine is quite rapid. Solubility problems with certain protein samples may be overcome by performing digestions in 6 M urea [467].

The varying specificities displayed by enzymes A and B toward individual C-terminal amino acids, as well as to penultimate residues, presents difficulties with respect to interpretation of the kinetic data thus obtained. Ideally, the CPase would cleave all peptide bonds, including those of proline, at identical rates irrespective of the nature of the side chain. This ideal has been approximated to a considerable degree by a class of enzymes termed CPase C isolated from citrus peel [468], French beans [469], and numerous other plant sources [464]. The term CPase C originated from studies of the citrus enzyme, and this exopeptidase is commercially available from Henley Company, New Jersey. Unlike CPases A and B, most of the C enzymes are inactivated by diisopropylphosphorofluoridate and may therefore exhibit similarities in mechanism to the well-documented "serine" proteinases. Although the CPases C release glycine at a very slow rate, all amino acids, including proline, are liberated during the course of digestion.

One of the major drawbacks to the more widespread application of these broad-specificity CPases to C-terminal analysis has been the difficulty in removing traces of protease impurity which, if present, would limit severely the interpretation of results thus obtained. An important advance in this regard has been the recent publication by Hayashi, Moore, and Stein [470], which describes the large-scale preparation and application to C-terminal sequence studies of a CPase from yeast. This enzyme was designated CPase Y on the basis of its source, yeast, and to distinguish it from the aforementioned CPases C to which it is similar in many of its properties. CPase Y is a glycoprotein, molecular weight 61,000, which consists of a single polypeptide chain of 442 residues. The first steps of the preparation involve purification of an inactive precursor of the enzyme, which is subsequently activated, and the native CPase Y is subjected to final purification by chromatography on DEAE-Sephadex [470]. CPase Y shares many characteristics in common with the CPases C; it has a pH optimum near 5, it displays broad specificity for all the amino acids including proline, except that glycine and aspartate retard its activity, and it may be employed in solutions containing 6 M urea.

Although extensive documentation of applications of CPase Y is not yet available, some examples taken from the work of Hayashi et al. [470] serve to illustrate its promise as a tool for C-terminal sequence analysis. In Table 15 are given the results obtained by long-term digestions of various protein substrates with CPase Y. It is clear from these studies that degradation may proceed far into the polypeptide chain. In the case of glucagon, the appearance of aspartate, nine residues from the C terminus, abruptly retarded further hydrolysis, and glycine exerted a similar effect on digestion of reduced and carboxymethylated (RCM)-insulin B chain and RCM-ribonuclease. Proline residues did not constitute a barrier to the enzyme. The rates of liberation of amino acids from native ribonuclease by digestion with CPase Y in 6 M urea and from RCM-ribonuclease are compared in Figure 5. These data permit the positioning of seven amino acids at the C terminus of the enzyme, extending by several residues the sequence obtained with CPase A. Since CPase Y can be obtained in large quantities and completely free of contaminating proteases, it is a most promising method for establishing considerable lengths of C-terminal sequences. Hopefully, the enzyme will soon be available commercially.

4 CONCLUDING REMARKS

It is our hope that this chapter will serve to stimulate interest among a wider audience of organic chemists in regard to some of the problems of contemporary protein chemistry. Generally speaking, the chemical modification and covalent structural analysis of proteins continue to follow two separate trends, both with respect to methodology and to the ultimate goals of each discipline. On the one hand, chemical modification studies attempt to ascertain the structural and functional roles of particular amino acid side chains with minimal perturbation of the native structure. Elucidation of the amino acid sequence, on the other hand, involves as a first step the complete disruption of all elements of secondary and tertiary structure. The importance of chemical modification in this area of endeavor relates primarily to its use as a means of producing or influencing cleavages in the polypeptide chain.

TABLE 15   DIGESTION OF GLUCAGON, RCM–INSULIN B CHAIN, AND RCM–RIBONUCLEASE BY CARBOXYPEPTIDASE Y[a]

Glucagon: —Arg—Ala—Gln—Asp—Phe—Val—Gln—Trp—Leu—Met—Asn—Thr
         0   0   0.07 0.69 0.85   0.84 0.92 0.81 1.4[b] 0.75

RCM–Insulin B chain: CmCys—Gly—Glu—Arg—Gly—Phe—Phe—Tyr—Thr—Pro—Lys—Ala
    (0.04 0.04 0.10)[c] 1.42 0.88 1.0 1.0 0.80 1.0
    (0.04 0.14

RCM–RNase: CmCys—Glu—Gly—Asn—Pro—Tyr—Val—Pro—Val—His—Phe—Asp—Ala—Ser—Val
    (0 0.08 0.03)[c]   0.53   0.79   0.80 0.82 0.86 0.92 1.2[d] 2.66

[a] Data from Hayashi, Moore, and Stein [470] with permission. The substrates, at a concentration of about 0.5%, were digested at pH 5.5 at an enzyme/substrate ratio of 1:100. Glucagon and the B chain of insulin were incubated at 35°C for 2 hr; RCM-ribonuclease was held at 25°C for 15 hr. The release of amino acids are expressed as moles per mole of substrate.

[b] Asn + Gln.

[c] The rate of hydrolysis slowed at Gly in these two chains.

[d] Ser + Asn.

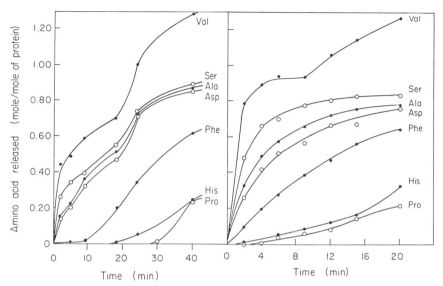

Figure 5 Comparison of exoproteolytic digestion of native ribonuclease A in 6 M urea (left) and reduced and carboxymethylated enzyme in the absence of urea (right) by yeast carboxypeptidase Y. Both mixtures at 25°C were buffered by 0.1 M pyridine acetate (pH 5.5) and contained 1% protein. The ratio of substrate to enzyme was 222 (left) and 280 (right). From Hayashi, Moore, and Stein [470] by permission.

Most of the reagents discussed above for the modification of individual amino acid side chains were designed to (1) increase the selectivity of the reaction; (2) introduce probes, the analytical properties of which facilitate quantitation; or (3) to react under conditions that better conserve the native conformation of the protein. Further development along these lines will certainly lead to improvements in these areas, especially in the selectivity now lacking both for the modification of certain amino acid side chains and for estimating the degree of exposure of classes of functional groups. At the present time little, if any, progress has been made in the interconversion of amino acid side chains, in the introduction of new functionalities of nonprotein nature, for example, nucleotides, carbohydrates, and lipids, or in the creation of new enzymes designed within the framework of existing protein structures. Thiosubtilisin [471] and dehydroalanine-chymotrypsin [472] still stand today as the only examples of interconversions of this kind in which active-site serine residues have been converted to cysteine and dehydroalanine, respectively, in the native enzymes. This novel and promising approach has not been extended as yet to the other functional groups of proteins, although the chemical methods are in the arsenal of the organic chemist. To date, the most productive work toward this end has been accomplished by chemical synthesis of altered polypeptides in order to assess the functional role of specific residues. The enzyme-catalyzed introduction of adenylate moieties into glutamine synthetase has been shown to be important relative to the mode of regulation of this enzyme (cf. [473]), whereas the functional role of the carbohydrate in a host of glycoproteins still remains obscure. Hopefully, the site-specific introduction by chemical means of substituents of this kind into existing proteins will provide new insights regarding their functional significance.

In considering the current status of protein covalent structural analysis, it is clearly apparent that there are many aspects of the field open to development or improvement by organic chemistry. The Edman degradation has been so valuable to studies of this kind that it is almost irreverent to suggest that it might be improved upon. However, if an alternative degradative scheme were devised in which the sensitivity to oxidation was minimized or eliminated, and in which milder conditions could be employed for sequential removal of the N-terminal residues, it might be possible to proceed much further into large peptides and proteins than is now possible. We have not discussed the many attempts that have been made to develop a useful chemical degradative method for the stepwise removal of amino acids from the C terminus of proteins, but this is not to minimize the importance of such a procedure. Chemical methods for opening the N-blocking pyrrolidonyl ring and the iminothiazolidinyl ring resulting from cleavage of S-cyanylated proteins [293] would find immediate application in the field, and the search continues for specific chemical cleavage procedures. Those of us whose work is intimately involved in these areas of protein chemistry look forward to the participation of a greater number of chemists in providing the answers to these and many other crucial questions.

REFERENCES

1. R. M. Heriott, *Advan. Protein Chem.*, 3, 169 (1947).
2. H. Fraenkel-Conrat, *Enzymes*, 1, 589 (1959).
3. C. A. Zittle, *J. Biol. Chem.*, 163, 111 (1946).
4. C. H. W. Hirs, *J. Biol. Chem.*, 235, 625 (1960).

5. C. H. W. Hirs, S. Moore, and W. H. Stein, *J. Biol. Chem.*, 235, 633 (1960).
6. D. H. Spackman, W. H. Stein, and S. Moore, *J. Biol. Chem.*, 235, 648 (1960).
7. D. G. Smyth, W. H. Stein, and S. Moore, *J. Biol. Chem.*, 238, 227 (1963).
8. J. T. Potts, A. Berger, J. Cooke, and C. B. Anfinsen, *J. Biol. Chem.*, 237, 1851 (1962).
9. S. Moore, D. H. Spackman, and W. H. Stein, *Anal. Chem.*, 30, 1185 (1958).
10. D. H. Spackman, W. H. Stein, and S. Moore, *Anal. Chem.*, 30, 1190 (1958).
11. S. Moore and W. H. Stein, *Methods Enzymol.*, 6, 819 (1963).
12. E. A. Barnard and W. D. Stein, *J. Mol. Biol.*, 1, 339, 350 (1959).
13. A. M. Crestfield, W. H. Stein, and S. Moore, *J. Biol. Chem.*, 238, 2413, 2421 (1963).
14. R. L. Heinrikson, W. H. Stein, A. M. Crestfield, and S. Moore, *J. Biol. Chem.*, 240, 2921 (1965).
15. M. O. Dayhoff, Ed., *Atlas of Protein Sequence and Structure*, Vol. 5, National Biomedical Research Foundation Silver Spring, Md., 1972.
16. W. A. Schroeder, *The Primary Structure of Proteins*, Harper and Row, New York, 1968.
17. S. B. Needleman, Ed., *Protein Sequence Determination*, Vol. 8, Springer-Verlag, Berlin, 1970.
18. C. H. W. Hirs, Ed., *Methods Enzymol.*, 11 (1967).
19. C. H. W. Hirs and S. N. Timasheff, Eds., *Methods Enzymol.*, 25, Part B (1972).
20. G. R. Stark, *Advan. Protein Chem.*, 24, 261 (1970).
21. T. F. Spande, B. Witkop, Y. Degani, and A. Patchornik, *Advan. Protein Chem.*, 24, 97 (1970).
22. A. F. S. A. Habeeb, in *Chemistry of the Cell Interface*, Part B, H. D. Brown, Ed., Academic Press, New York, 1971, p. 259.
23. L. Brand and J. R. Gohlke, *Ann. Rev. Biochem.*, 41, 843 (1972).
24. L. A. Cohen, *Ann. Rev. Biochem.*, 37, 695 (1968).
25. A. N. Glazer, *Ann. Rev. Biochem.*, 39, 101 (1970).
26. G. A. Means, and R. E. Feeney, *Chemical Modification of Proteins*, Holden-Day, San Francisco, 1971.
27. B. L. Vallee and J. F. Riordan, *Ann. Rev. Biochem.*, 38, 733 (1969).
28. J. F. Riordan and M. Sokolovsky, *Accounts Chem. Res.*, 4, 353 (1971).
29. S. J. Singer, *Advan. Protein Chem.*, 22, 1 (1967).
30. B. R. Baker, *Design of Active-Site-Directed Irreversible Enzyme Inhibitors*, Wiley, New York, 1967.
31. A. Marglin and R. B. Merrifield, *Ann. Rev. Biochem.*, 39, 841 (1970).
32. M. C. Lin, B. Gutte, D. G. Caldi, S. Moore, and R. B. Merrifield, *J. Biol. Chem.*, 247, 4768 (1972).
33. G. R. Sanchez, I. M. Chaiken, and C. B. Anfinsen, *J. Biol. Chem.*, 248, 3653 (1973).
34. R. E. Dickerson, *Ann. Rev. Biochem.*, 41, 815 (1972).

35. G. P. Hess and J. A. Rupley, *Ann. Rev. Biochem.*, **40**, 1013 (1971).
36. A. M. Scanu and C. Wisdom, *Ann. Rev. Biochem.*, **41**, 703 (1972).
37. R. D. Marshall, *Ann. Rev. Biochem.*, **41**, 673 (1972).
38. J. A. Rupley and A. F. Shrake, private communication.
39. M. O. Dayhoff and L. T. Hunt, Eds., *Atlas of Protein Sequence and Structure*, Vol. 5, National Biomedical Research Foundation, Silver Spring, Md., 1972, D-355.
40. J. Steinhardt and S. Beychok, in *The Proteins*, H. Neurath, Ed., Vol. 2, Academic Press, New York, 1964, p. 140.
41. R. Taft, in *Steric Effects in Organic Chemistry*, M. S. Newman, Ed., Wiley, New York, Chap. 13, 1956.
42. M. Friedman and J. S. Wall, *J. Am. Chem. Soc.*, **86**, 3735 (1964).
43. M. Friedman and J. S. Wall, *J. Org. Chem.*, **31**, 2888 (1966).
44. R. Fields, *Methods Enzymol.*, **25B**, 464 (1972).
45. J. F. Riordan and B. L. Vallee, *Methods Enzymol.*, **25B**, 494 (1972).
46. H. Kaplan, K. J. Stevenson, and B. S. Hartley, *Biochem. J.*, **124**, 289 (1971).
47. H. Kaplan, *J. Mol. Biol.*, **72**, 153 (1972).
48. M. Irie, T. Mujasaka, and K. Arakawa, *J. Biochem.* (Tokyo), **72**, 65 (1972).
49. M. H. Klapper and I. M. Klotz, *Methods Enzymol.*, **25B**, 531 (1972).
50. P. J. G. Butler and B. S. Hartley, *Methods Enzymol.*, **25B**, 191 (1972).
51. M. Z. Atassi and A. F. S. A. Habeeb, *Methods Enzymol.*, **25B**, 546 (1972).
52. F. H. White, *Methods Enzymol.*, **25B**, 541 (1972).
53. A. F. S. A. Habeeb, *Methods Enzymol.*, **25B**, 558 (1972).
54. M. J. Hunter and M. L. Ludwig, *Methods Enzymol.*, **25B**, 585 (1972).
55. F. Wold, *Methods Enzymol.*, **25B**, 623 (1972).
56. M. O'Leary and G. Samberg, *J. Am. Chem. Soc.*, **93**, 3530 (1971).
57. R. B. Scheele and M. A. Laufber, *Biochemistry*, **8**, 3597 (1969).
58. L. Rao and T. Hofmann, *Can. J. Biochem.*, **48**, 1249 (1970).
59. R. A. Goldfarb, *Biochim. Biophys. Acta*, **200**, 1 (1970).
60. P. F. Hollenberg, M. Flashner, and M. J. Coon, *J. Biol. Chem.*, **246**, 946 (1971).
61. P. Roschlau and B. Hess, *Hoppe-Seyler's Z. Physiol. Chem.*, **353**, 944 (1972).
62. R. Fields, *Biochem. J.*, **124**, 581 (1971).
63. C. Clark, and K. Yielding, *Arch. Biochem. Biophys.*, **143**, 158 (1971).
64. C. J. Coffee, R. A. Bradshaw, B. R. Goldin, and C. Frieden, *Biochemistry*, **10**, 3516 (1971).
65. B. R. Goldin and C. Frieden, *Biochemistry*, **10**, 3527 (1971).
66. J. Forstner and J. Manery, *Biochem. J.*, **125**, 343 (1971).
67. A. Gertler, *Eur. J. Biochem.*, **23**, 36 (1971).
68. P. Valenzuela and M. L. Bender, *Biochim. Biophys. Acta*, **250**, 538 (1971).
69. R. Zschocke, M. Chiao, and A. Bezkorovainy, *Eur. J. Biochem.*, **27**, 145 (1972).

70. R. L. Heinrikson, R. Sterner, C. Noyes, B. S. Cooperman, and R. H. Bruckmann, *J. Biol. Chem.*, **248**, 2521 (1973).
71. D. D. F. Shiao, R. Lumry, and S. Rajender, *Eur. J. Biochem.*, **29**, 377 (1972).
72. Y. Nakagawa, S. Capetillo, and B. Jirgensons, *J. Biol. Chem.*, **247**, 5703 (1972).
73. A. F. S. A. Habeeb and M. Z. Atassi, *Biochem.*, **9**, 4939 (1970).
74. Y. Nakagawa and G. Perlmann, *Arch. Biochem. Biophys.*, **149**, 476 (1972).
75. Y. Ohta, H. Nakamura, and T. Samejima, *J. Biochem.* (Tokyo), **72**, 521 (1972).
76. P. A. Kendall and E. A. Barnard, *Biochim. Biophys. Acta*, **188**, 10 (1969).
77. D. Cechova and G. Maszynska, *FEBS Letters*, **8**, 274 (1970).
78. L. Spero, H. M. Jacoby, J. E. Dalidowicz, and S. J. Silverman, *Biochim. Biophys. Acta*, **251**, 345 (1971).
79. B. V. Plapp, S. Moore, and W. H. Stein, *J. Biol. Chem.*, **246**, 939 (1971).
80. E. Karlsson, D. Eaker, and G. Ponterius, *Biochim. Biophys. Acta*, **257**, 235 (1972).
81. D. Kowalski and M. Laskowski, Jr., *Biochemistry*, **11**, 3451 (1972).
82. D. M. Glick and E. A. Barnard, *Biochim. Biophys. Acta*, **214**, 326 (1970).
83. A. Nareddin and T. Inagami, *Biochem. Biophys. Res. Commun.*, **36**, 999 (1969).
84. N. C. Robinson, H. Neurath, and K. Walsh, *Biochemistry*, **12**, 414, 420 (1973).
85. J. Abita, M. Lazdunski, P. Bonsen, W. Pieterson, and G. DeHaas, *Eur. J. Biochem.*, **30**, 37 (1972).
86. B. V. Plapp, *J. Biol. Chem.*, **245**, 1727 (1970).
87. F. M. Veronese, D. Piszkiewicz, and E. L. Smith, *J. Biol. Chem.*, **247**, 754 (1972).
88. T. Chulkova and V. Orekhovich, *Biokhimyia*, **35**, 237 (1970).
89. S. Magnusson and T. Hofmann, *Can. J. Biochem.*, **48**, 432 (1970).
90. A. Kurosky and T. Hofmann, *Can. J. Biochem.*, **50**, 1282 (1972).
91. J. Dixon and T. Hofmann, *Can. J. Biochem.*, **48**, 671 (1970).
92. A. Kurosky, J. Graham, J. Dixon, and T. Hofmann, *Can. J. Biochem.*, **49**, 529 (1971).
93. K. L. Carraway and D. E. Koshland, *Methods Enzymol.*, **25B**, 616 (1973).
94. H. G. Khorana, *Chem. Rev.*, **53**, 145 (1953).
95. A. F. Hegarty and T. J. Bruice, *J. Am. Chem. Soc.*, **92**, 6568 (1970).
96. K. L. Carraway and R. B. Triplett, *Biochim. Biophys. Acta*, **200**, 564 (1970).
97. K. L. Carraway and D. E. Koshland, *Biochim. Biophys. Acta*, **160**, 272 (1968).
98. K. J. Kramer and J. A. Rupley, *Arch. Biochem. Biophys.*, **156**, 414 (1973).
99. T. Imoto, L. N. Johnson, A. C. T. North, D. C. Phillips, and J. A. Rupley, *The Enzymes*, P. D. Boyer, Ed., Vol. 7, Academic Press, New York, 1972, p. 665.

100. J. F. Riordan and H. Hayshida, *Biochem. Biophys. Res. Commun.*, **41**, 122 (1970).
101. P. Bodlaender, G. Feinstein, and E. Shaw, *Biochemistry*, **8**, 4941 (1969).
102. G. Feinstein, P. Bodlaender, and E. Shaw, *Biochemistry*, **8**, 4949 (1969).
103. P. H. Petra, *Biochemistry*, **10**, 3163 (1971).
104. P. H. Petra and H. Neurath, *Biochemistry*, **10**, 3171 (1971).
105. P. E. Wilcox, *Methods Enzymol.*, **25B**, 596 (1972).
106. S. M. Parsons, L. Jao, F. W. Dahlquist, C. L. Borders, T. Groff, J. Racs, and M. A. Raftery, *Biochemistry*, **8**, 700 (1969).
107. H. Nakayama, K. Tanizawa, and Y. Kanaoka, *Biochem. Biophys. Res. Commun.*, **40**, 537 (1970).
108. M. Z. Atassi, *Biochim. Biophys. Acta*, **303**, 379 (1973).
109. W. Robinson and B. Belleau, *J. Am. Chem. Soc.*, **94**, 4376 (1972).
110. F. S. Chu and E. Crary, *Biochim. Biophys. Acta*, **194**, 287 (1969).
111. J. P. Abita and M. Lazdunski, *Biochem. Biophys. Res. Commun.*, **35**, 707 (1969).
112. J. P. Abita, S. Maroux, M. Delaage, and M. Lazdunski, *FEBS Letters*, **4**, 203 (1969).
113. K. L. Carraway, P. Spoerl, and D. E. Koshland, *J. Mol. Biol.*, **42**, 133 (1969).
114. A. Bezkorovainy and D. Grohlich, *Biochim. Biophys. Acta*, **214**, 37 (1970).
115. A. W. Eyl and T. Inagami, *J. Biol. Chem.*, **246**, 738 (1971).
116. H. Ozawa, *Biochemistry*, **9**, 2158 (1970).
117. R. W. Colman, *J. Biol. Chem.*, **246**, 4497 (1971).
118. C. C. Chang, C. C. Yang, K. Nakai, and K. Hayashi, *Biochim. Biophys. Acta*, **251**, 334 (1971).
119. R. Frater, *FEBS Letters*, **12**, 186 (1971).
120. F. Bossa, D. Barra, P. Vecchini, and C. Turano, *Emzymologia*, **40**, 24 (1971).
121. A. Z. Budzynski and G. E. Means, *Biochim. Biophys. Acta*, **236**, 767 (1971).
122. K. R. Adams, *Comp. Rend. Lab. Carlsberg*, **38**, 481 (1972).
123. M. Z. Atassi and R. P. Singhal, *J. Biol. Chem.*, **247**, 5980 (1972).
124. L. Matyash, O. Ogloblina, and V. Stepanov, *Biokhimiya*, **37**, 1067 (1972).
125. N. R. Davis and T. E. Walker, *Biochem. Biophys. Res. Commun.*, **48**, 1656 (1972).
126. R. Epand and R. Epand, *Biochim. Biophys. Acta*, **285**, 176 (1972).
127. G. Klipperstein, *Biochem. Biophys. Res. Commun.*, **49**, 1474 (1972).
128. A. R. Fersht and J. Sperling, *J. Mol. Biol.*, **74**, 137 (1973).
129. S. D. Lewis and J. A. Shafer, *Biochim. Biophys. Acta*, **303**, 284 (1973).
130. D. V. Godin and S. L. Schrier, *Biochemistry*, **9**, 4068 (1970).
131. A. Gertler, *FEBS Letters*, **19**, 255 (1971).
132. T. Valveva and L. Girodman, *Biokhimiya*, **35**, 732 (1970).
133. H. Keilova, *FEBS Letters*, **6**, 312 (1970).

134. J. Kay and A. Ryle, *Biochem. J.*, **123**, 75 (1971).
135. P. Meitner, *Biochem. J.*, **124**, 673 (1971).
136. H. Keilova and C. Lapresle, *FEBS Letters*, **9**, 348 (1972).
137. K. Takahashi, F. Mizobe, and W. Chang, *J. Biochem.* (Tokyo), **71**, 161 (1972).
138. F. Mizobe, K. Takahashi, and T. Ando, *J. Biochem.* (Tokyo), **73**, 61 (1973).
139. S. Husain, J. Fergason, J. Fruton, *Proc. Nat. Acad. Sci. U.S.*, **68**, 2765 (1971).
140. A. Paterson and J. Knowles, *Eur. J. Biochem.*, **31**, 510 (1972).
141. J. A. Yankeelov, *Methods Enzymol.*, **25B**, 566 (1972).
142. K. Takahashi, *J. Biol. Chem.*, **243**, 6171 (1968).
143. N. V. Chuyen, T. Kurata, and M. Fujimaki, *Agr. Biol. Chem.*, **37**, 327 (1973).
144. K. Nakaya, T. Suzuki, O. Takenaka, and K. Shibata, *Biochim. Biophys, Acta*, **194**, 301 (1969).
145. T. Fukui, A. Kamogawa, and Z. Nikuni, *J. Biochem.* (Tokyo), **67**, 211 (1970).
146. K. Takahashi, *J. Biochem.* (Tokyo), **68**, 659 (1970).
147. J. E. Delarco and I. E. Liener, *Biochim. Biophys. Acta*, **303**, 274 (1973).
148. M. M. Werber and M. Sokolovsky, *Biochem. Biophys. Res. Commun.*, **48**, 384 (1972).
149. B. Keil, *FEBS Letters*, **14**, 181 (1971).
150. J. Berghauser and I. Falderbaum, *Hoppe-Seyler's Z. Physiol. Chem.*, **352**, 1189 (1971).
151. J. A. Yankeelov, *Biochemistry*, **9**, 2433 (1970).
152. J. A. Yankeelov and D. Aoree, *Biochem. Biophys. Res. Commun.*, **42**, 886 (1971).
153. P. C. Yang and G. W. Schwert, *Biochemistry*, **11**, 2218 (1972).
154. A. Signor, G. Bonora, L. Biordi, D. Nisato, A. Marzotto, and E. Scoffore, *Biochemistry*, **10**, 2748 (1971).
155. G. Perlman and L. Lorand, *Methods Enzymol.*, **19**, Sections I and II (1971).
156. T. E. Banks, B. K. Blossey, and J. A. Shafer, *J. Biol. Chem.*, **244**, 6323 (1969).
157. W. Brown and F. Wold, *Biochemistry*, **12**, 828 (1973).
158. W. Brown and F. Wold, *Biochemistry*, **12**, 835 (1973).
159. K. Iwai and T. Ando, *Methods Enzymol.*, **11**, 263 (1967).
160. D. Levy and F. H. Carpenter, *Biochemistry*, **9**, 3215 (1970).
161. T. C. Bruice and S. Benkovic, "Bioorganic Mechanism," Vol. I., W. A. Benjamin, New York, 1966, p. 125.
162. W. B. Melchior and D. Fahrney, *Biochemistry*, **9**, 251 (1970).
163. L. Pradel and R. Kassab, *Biochim. Biophys. Acta*, **167**, 317 (1968).
164. M. Wells, *Biochemistry*, **12**, 1086 (1973).
165. Y. Nakagawa and M. L. Bender, *Biochemistry*, **9**, 259 (1970).

166. R. L. Heinrikson, *J. Biol. Chem.*, **246**, 4090 (1971).
167. J. Bello and E. Nowoswiat, *Eur. J. Biochem.*, **22**, 225 (1971).
168. F. R. N. Gurd, *Methods Enzymol.*, **25B**, 424 (1972).
169. R. Oshima and P. A. Price, *J. Biol. Chem.*, **248**, 7522 (1973).
170. J. F. Riordan and B. C. Vallee, *Methods Enzymol.*, **25B**, 521 (1972).
171. C. Huc, A. Olomucki, Le-Thi-Lan, D. Pho, and N. Van Thoai, *Eur. J. Biochem.*, **21**, 161 (1971).
172. F. Thomé-Beau, Le-Thi-Lan, A. Olomucki, and N. Van Thoai, *Eur. J. Biochem.*, **19**, 270 (1971).
173. D. L. Morris and J. McKinley-McKee, *Eur. J. Biochem.*, **29**, 515 (1972).
174. U. Nylén and G. Pettersson, *Eur. J. Biochem.*, **27**, 578 (1972).
175. M. Rippa, M. Signorini, and Pontremoli, *Arch. Biochem. Biophys.*, **150**, 503 (1972).
176. N. Tudball, R. Bailey-Wood, and P. Thoma, *Biochem. J.*, **129**, 419 (1972).
177. R. B. Wallis and J. J. Holbrook, *Biochem. J.*, **133**, 183 (1973).
178. G. Gachelio, L. Goldstein, D. Hofrung, and A. J. Kalb, *Eur. J. Biochem.*, **30**, 155 (1972).
179. R. M. Epand, R. F. Epand, and V. Grey, *Arch. Biochem. Biophys.*, **154**, 132 (1973).
180. F. J. Costellino and R. L. Hill, *J. Biol. Chem.*, **245**, 417 (1970).
181. I. Covelli, L. Frati, and J. Wolff, *J. Biochem.*, **12**, 1043 (1973).
182. B. Anderton, *Eur. J. Biochem.*, **15**, 562, 568 (1970).
183. T. E. Hugli and F. R. N. Gurd, *J. Biol. Chem.*, **245**, 1930 (1970).
184. T. E. Hugli and F. R. N. Gurd, *J. Biol. Chem.*, **245**, 1939 (1970).
185. P. Whitney, *Eur. J. Biochem.*, **16**, 126 (1970).
186. K. Ohtsuki, N. Kaziwara, and H. Hatano, *J. Biochem.* (Tokyo), **68**, 137 (1970).
187. A. Takenaka, O. Takenaka, H. Horinishi, and K. Shibata, *J. Biochem.* (Tokyo), **67**, 397 (1970).
188. R. Apitz-Castro and Z. Svarez, *Biochim. Biophys. Acta*, **198**, 176 (1970).
189. F. Zaheer and B. H. Nicholson, *Biochim. Biophys. Acta*, **251**, 38 (1971).
190. W. H. Cruickshank and H. Kaplan, *Biochem. Biophys. Res. Commun.*, **46**, 2134 (1972).
191. H. R. Horton and D. E. Koshland, *Methods Enzymol.*, **25B**, 468 (1972).
192. G. M. Loudon and D. E. Koshland, *J. Biol. Chem.*, **245**, 2247 (1970).
193. C. P. Heinrich, S. Adam, and W. Arnold, *FEBS Letters*, **33**, 181 (1973).
194. A. Fontana and E. Scoffone, *Methods Enzymol.*, **25B**, 482 (1972).
195. A. Previero, M. A. Coletti-Previero, and J. C. Cavadore, *Biochim. Biophys. Acta*, **147**, 453 (1967).
196. K. A. Schellenberg and G. W. McLean, *Biochim. Biophys. Acta*, **191**, 727 (1969).
197. C. Chang and K. Hayashi, *Biochem. Biophys. Res. Commun.*, **37**, 841 (1969).
198. K. Takahashi, *J. Biochem.* (Tokyo), **67**, 541 (1970).
199. M. Irie, *J. Biochem.* (Tokyo), **68**, 31 (1970).

200. A. Seto, S. Sato, and N. Tamiya, *Biochim. Biophys. Acta*, 214, 483 (1970).
201. T. Barman, *J. Mol. Biol.*, 52, 391 (1970).
202. L. Chao and E. R. Einstein, *J. Biol. Chem.*, 245, 6397 (1970).
203. T. L. Poulos and P. A. Price, *J. Biol. Chem.*, 246, 4041 (1971).
204. Y. Baba, A. Arimura, and A. Schally, *Biochem. Biophys. Res. Commun.*, 45, 483 (1971).
205. M. Irie and K. Harada, *J. Biochem.* (Tokyo), 72, 1351 (1972).
206. H. Bachmayer, *FEBS Letters*, 23, 217 (1972).
207. V. Keil and B. Keil, *J. Mol. Biol.*, 67, 495 (1972).
208. A. Ford-Hutchinson and D. Perkins, *Eur. J. Biochem.*, 25, 419 (1972).
209. C. P. Heinrich, K. Noach, and D. Wiss, *Biochem. Biophys. Res. Commun.*, 49, 1427 (1972).
210. T. Barman, *Biochim. Biophys. Acta*, 258, 297 (1972).
211. T. Barman and W. Bagshaw, *Biochim. Biophys. Acta*, 278, 491 (1972).
212. B. Keil, *FEBS Letters*, 29, 25 (1973).
213. S. Lindskog and A. Nilsson, *Biochim. Biophys. Acta*, 295, 117 (1973).
214. I. Parikh, I and G. S. Omenn, *Biochemistry*, 10, 1173 (1971).
215. Y. Shechter, Y. Burstein, and A. Patchornik, *Biochemistry*, 11, 653 (1972).
216. I. Aviram and A. Schejter, *Biochim. Biophys. Acta*, 229, 113 (1971).
217. A. Holmgren, *Eur. J. Biochem.*, 26, 528 (1972).
218. J. Riordan and B. L. Vallee, *Methods Enzymol.*, 25B, 515 (1972).
219. J. Williams and J. Lowe, *Biochem. J.*, 121, 203 (1971).
220. J. Williams and J. Lowe, *Biochem. J.*, 121, 500 (1972).
221. M. Sokolovsky and J. F. Riordan, *FEBS Letters*, 9, 239 (1970).
222. J. F. Riordan and B. L. Vallee, *Methods Enzymol.*, 25B, 500 (1972).
223. J. F. Riordan and B. L. Vallee, *Biochemistry*, 3, 1768 (1964).
224. M. J. Gorbunoff, *Methods Enzymol.*, 25B, 506 (1972).
225. M. J. Gorbunoff, *Biopolymers*, 11, 2233 (1972).
226. O. A. Roholt and D. Pressman, *Methods Enzymol.*, 25B, 438 (1972).
227. J. Vincent, M. Lazdunski, and M. Delaage, *Eur. J. Biochem.*, 12, 250 (1970).
228. R. Boesel and F. Carpenter, *Biochem. Biophys. Res. Commun.*, 38, 678 (1970).
229. R. Kassab, A. Faltoren, and L. Pradel, *Eur. J. Biochem.*, 12, 264 (1970).
230. C. Szpirer and R. Jeener, *FEBS Letters*, 8, 229 (1970).
231. D. Tsuru, T. Yoshida, and J. Fukumoto, *J. Biochem.* (Tokyo), 67, 867 (1970).
232. J. M. Connellan and D. C. Shaw, *J. Biol. Chem.*, 245, 2845 (1970).
233. W. Domschke, C. V. Hinueber, and G. F. Domagk, *Biochim. Biophys. Acta*, 207, 485 (1970).
234. J. L. York and C. C. Fan, *Biochemistry*, 10, 1659 (1971).
235. P. Christen and J. F. Riordan, *Biochemistry*, 9, 3025 (1970).
236. L. Ma, J. Brovetto-Cruz, and C. H. Li, *Biochemistry*, 9, 2302 (1970).

237. M. Sokolovsky, I. Aviram, and A. Schejter, *Biochemistry*, 9, 5113 (1970).
238. A. Schejter, I. Aviram, and M. Sokolovsky, *Biochemistry*, 9, 5118 (1970).
239. P. G. Malan and H. Edelhoch, *Biochemistry*, 9, 3205 (1970).
240. N. Boyd and D. Smith, *Can. J. Biochemistry*, 49, 154 (1971).
241. F. Dorner, *J. Biol. Chem.*, 246 5896 (1971).
242. N. S. Otchin and H. Metzger, *J. Biol. Chem.*, 246, 7051 (1971).
243. R. Lundblad and J. Harrison, *Biochem. Biophys. Res. Commun.*, 45, 1344 (1971).
244. H. Gattner, *Hoppe-Seyler's Z. Physiol. Chem.*, 352, 7 (1971).
245. A. Strosberg, B. Van Hoeck, and L. Kanarek, *Eur. J. Biochem.*, 19, 36 (1971).
246. T. E. Hugli and W. H. Stein, *J. Biol. Chem.*, 246, 7191 (1971).
247. D. Piszkiewicz, M. Landon, and E. L. Smith, *J. Biol. Chem.*, 246, 1324 (1971).
248. A. Waheed and S. Skall, *Biochim. Biophys. Acta*, 242, 172 (1971).
249. C. Chang, C. Ganz, K. Hamaguchi, K. Nakai, and K. Hayashi, *Biochim. Biophys. Acta*, 236, 164 (1971).
250. K. Goto, N. Takahashi, and T. Murachi, *J. Biochem.* (Tokyo), 70, 157 (1971).
251. W. L. Denton and K. E. Ebner, *J. Biol. Chem.*, 246, 4053 (1971).
252. P. Christen, B. L. Vallee, and R. T. Simpson, *Biochemistry*, 10, 1377 (1971).
253. M. K. Dabbous, M. Seif, and E. C. Brinkley, *Biochem. Biophys. Res. Commun.*, 48, 1586 (1972).
254. Y. P. Lui and R. E. Handschumacher, *J. Biol. Chem.*, 247, 66 (1972).
255. M. Sokolovsky, *Eur. J. Biochem.*, 25, 267 (1972).
256. K. Cheng and J. Piecer, *J. Biol. Chem.*, 247, 7163 (1972).
257. M. Sairam, H. Papkoff, and C. H. Li, *Biochim. Biophys. Acta*, 278, 421 (1972).
258. O. Polyanovsky, T. Demidkina, and C. Egorov, *FEBS Letters*, 23, 262 (1972).
259. M. L. Raymond and A. T. Tu, *Biochim. Biophys. Acta*, 285, 498 (1972).
260. S. Shifrin and B. G. Solis, *J. Biol. Chem.*, 247, 4121 (1972).
261. R. Chicheportiche, C. Rochat, F. Sampieri, and M. Lazdunski, *Biochemistry*, 11, 168 (1972).
262. S. E. Papaioannou and I. E. Liener, *J. Biol. Chem.*, 245, 4931 (1970).
263. S. Wilk, A. Meister, and R. H. Haschemeyer, *Biochemistry*, 9, 2039 (1970).
264. T. Azuma, K. Hamaguchi, and S. Migita, *J. Biochem.* (Tokyo), 69, 535 (1971).
265. C. C. Fan and J. L. York, *Biochem. Biophys. Res. Commun.*, 47, 472 (1972).
266. I. Sjoholm, A. Eilenas, and J. Sjoquist, *Eur. J. Biochem.*, 29, 455 (1972).
267. Y. Y. Nakae and K. Hamaguchi, *J. Biochem.* (Tokyo), 72, 1155 (1972).
268. R. L. Lundblad, J. H. Harrison, and K. G. Mann, *Biochemistry*, 12, 409 (1973).

269. M. J. Gorbunoff, *Biochemistry*, 10, 250 (1971).
270. M. J. Gorbunoff, *Arch. Biochem. Biophys.*, 138, 684 (1970).
271. M. J. Gorbunoff and S. Timasheff, *Arch. Biochem. Biophys.*, 152, 413 (1972).
272. J. Thomas and J. Harris, *Biochem. J.*, 119, 307 (1970).
273. B. Seon, O. A. Roholt, and D. Pressman, *Biochim. Biophys. Acta*, 200, 81 (1970).
274. B. Seon, O. A. Roholt, and D. Pressman, *J. Biol. Chem.*, 246, 887 (1971).
275. S. Libor and P. Elodi, *Eur. J. Biochem.*, 12, 336 (1970).
276. S. Libor and P. Elodi, *Eur. J. Biochem.*, 12, 345 (1970).
277. E. B. McGowan and E. Stellwagen, *Biochemistry*, 9, 3047 (1970).
278. V. Csányi, I. Ferencz, and I. Mile, *Biochim. Biophys. Acta*, 236, 619 (1971).
279. O. Roholt and D. Pressman, *Eur. J. Biochem.*, 18, 79 (1971).
280. A. Fattoum, R. Kassab, and L. Pradel, *Eur. J. Biochem.*, 22, 445 (1971).
281. C. Garratt, D. Harrison, and M. Wicks, *Biochem. J.*, 126, 123 (1972).
282. I. Vasiletz, M. Shavlovsky, and S. Neifakh, *Eur. J. Biochem.*, 25, 498 (1972).
283. C. Rochat, F. Sampieri, H. Rochat, and F. Miranda, *Biochimie*, 54, 445 (1972).
284. T. E. Huntley and P. Strittmatter, *J. Biol. Chem.*, 247, 4648 (1972).
285. Reference 187.
286. R. W. Furlanetto and E. T. Kaiser, *J. Am. Chem. Soc.*, 92, 6980 (1970).
287. J. F. Riordan and B. L. Vallee, *Methods Enzymol.*, 25B, 449 (1972).
288. A. F. S. A. Habeeb, *Methods Enzymol.*, 25B, 457 (1972).
289. J. Meienhofer, J. Czombos, and H. Maeda, *J. Am. Chem. Soc.*, 93, 3080 (1971).
290. M. Friedman, L. H. Krull, and J. F. Cavins, *J. Biol. Chem.*, 245, 3868 (1970).
291. M. A. Hermodson, L. H. Ericsson, K. Titani, H. Neurath, and K. A. Walsh, *Biochemistry*, 11, 4493 (1972).
292. Y. Degani and A. Patchornik, *J. Org. Chem.*, 36, 2727 (1971).
293. G. R. Jacobson, M. H. Schaffer, G. R. Stark, and T. C. Vanaman, *J. Biol. Chem.*, 248, 6583 (1973).
294. H. A. Itano and E. A. Robinson, *J. Biol. Chem.*, 247, 4819 (1972).
295. K. Kuromizu and J. Meienhofer, *J. Biol. Chem.*, 247, 5646 (1972).
296. M. S. Masri, J. J. Windle, and M. Friedman, *Biochem. Biophys. Res. Commun.*, 47, 1408 (1972).
297. A. Fenselau and P. Weigel, *Biochim. Biophys. Acta*, 198, 192 (1970).
298. S. Husain and G. Lowe, *Biochem. J.*, 117, 291 (1970).
299. C. Reynolds and J. McKinley-McKee, *Biochem. J.*, 119, 801 (1970).
300. A. M. Scanu, *Biochim. Biophys. Acta*, 200, 570 (1970).
301. A. Light, B. C. Hardwick, L. M. Hatfield, and D. Sondack, *J. Biol. Chem.*, 244, 6289 (1969).
302. D. L. Sondack and A. Light, *J. Biol. Chem.*, 246, 1630 (1971).

303. L. M. Hatfield, S. K. Banerjee, and A. Light, *J. Biol. Chem.*, 246, 6303 (1971).
304. J. Bridge, *Biochem. J.*, 126, 21 (1972).
305. E. M. Gregory and J. H. Harrison, *Biochem. Biophys. Res. Commun.*, 40, 995 (1970).
306. R. Atherton, J. Laws, and A. Thomas, *Biochem. J.*, 118, 903 (1970).
307. L. H. M. Janssen, S. H. De Bruin, and G. A. J. Van Os, *Biochim. Biophys. Acta*, 236, 777 (1971).
308. M. Brunori, B. Talbot, A. Colosimo, E. Antonini, and J. Wyman, *J. Mol. Biol.*, 65, 423 (1972).
309. W. Birchmeier, B. Glatthan, K. Winterhalter, and R. Bradshaw, *Eur. J. Biochem.*, 28, 533 (1972).
310. F. Dickenson, *Biochem. J.*, 126, 133 (1972).
311. B. Szajáni, M. Sajgó, E. Biszku, P. Friedrich, and G. Szabocsi, *Eur. J. Biochem.*, 15, 171 (1970).
312. Y. Morino and M. Okamoto, *Biochem. Biophys. Res. Commun.*, 40, 600 (1970).
313. S. Husain and G. Lowe, *Biochem. J.*, 117, 333 (1970).
314. S. Husain and G. Lowe, *Biochem. J.*, 117, 341 (1970).
315. J. Moore and A. Fenselau, *Biochemistry*, 11, 3752 (1972).
316. J. Moore and A. Fenselau, *Biochemistry*, 11, 3762 (1972).
317. Y. Lue, R. Kaber, I. Norton, and F. Hartman, *Biochem. Biophys. Res. Commun.*, 45, 34 (1971).
318. J. Holbrook and R. Stinson, *Biochem. J.*, 120, 289 (1970).
319. C. Roustan, E. Terrossian, and L. Pradel, *Eur. J. Biochem.*, 17, 467 (1970).
320. S. Simon, D. Arndt, and W. Konigsberg, *J. Mol. Biol.*, 58, 69 (1971).
321. J. Moffat, *J. Mol. Biol.*, 58, 79 (1971).
322. J. Moffat, S. Simon, and W. Konigsberg, *J. Mol. Biol.*, 58, 89 (1971).
323. E. M. Gregory, F. J. Yost, M. S. Rohrbach, and J. H. Harrison, *J. Biol. Chem.*, 246, 5491 (1971).
324. W. B. Freedberg and J. K. Hardman, *J. Biol. Chem.*, 246, 1439 (1971).
325. W. B. Freedberg and J. K. Hardman, *J. Biol. Chem.*, 246, 1449 (1971).
326. D. J. Arndt and W. Konigsberg, *J. Biol. Chem.*, 246, 2594 (1971).
327. D. J. Arndt, S. R. Simon, T. Maita, and W. Konigsberg, *J. Biol. Chem.*, 246, 2602 (1971).
328. H. Teraoka, S. Naito, K. Izui, and H. Katsuki, *J. Biochem.* (Tokyo), 71, 157 (1972).
329. G. B. Ralston, *Comp. Rend. Trav. Lab. Carlsberg*, 38, 499 (1972).
330. N. T. Felberg and T. C. Hollocher, *J. Biol. Chem.*, 247, 4539 (1972).
331. R. Lee, W. D. McElroy, *Biochemistry*, 8, 130 (1969).
332. W. H. Porter, L. W. Cunningham, and W. M. Mitchell, *J. Biol. Chem.*, 246, 7675 (1971).
333. R. F. Colman, *Biochemistry*, 8, 888 (1969).
334. P. M. Wassarman and J. P. Major, *Biochemistry*, 8, 1076 (1969).
335. K. Kleppe and S. Damjanovich, *Biochim. Biophys. Acta*, 185, 88 (1969).

336. R. Verger, L. Sarda, and P. Desnuelle, *Biochim. Biophys. Acta*, 207, 377 (1970).
337. T. C. Vanaman and G. R. Stark, *J. Biol. Chem.*, 245, 3565 (1970).
338. I. Cournil and M. Arrio-Dupont, *Biochem. Biophys. Res. Commun.*, 43, 40 (1971).
339. D. Demé, O. Trautmann, and F. Chantogner, *Eur. J. Biochem.*, 20, 269 (1971).
340. K. Nagami, *Biochem. Biophys. Res. Commun.*, 47, 803 (1972).
341. K. Brocklehurst, M. Kierstan, and G. Little, *Biochem. J.*, 128, 811 (1972).
342. R. Jones, R. Dwek, and I. Walker, *FEBS Letters*, 26, 92 (1972).
343. F. Dolder, *Biochim. Bioplys. Acta*, 207, 286 (1970).
344. K. Brocklehurst, E. Crook, and M. Kierstan, *Biochem. J.*, 128, 979 (1972).
345. J. Twu, and F. Wold, *Biochemistry*, 12, 381 (1973).
346. R. F. Colman, *Biochim. Biophys. Acta*, 191, 469 (1969).
347. D. J. Edwards, R. L. Heinrikson, and A. E. Chung, *Biochemistry*, 13, 677 (1974).
348. J. Kraut, in *The Enzymes*, Vol. 3, P. D. Boyer, Ed., 3rd ed., Academic Press, New York, 1971, p. 165.
349. R. P. Taylor, J. B. Vatz, and R. Lumry, *Biochemistry*, 12, 2933 (1973).
350. F. S. Chu and M. S. Bergdoll, *Biochim. Biophys. Acta*, 194, 279 (1969).
351. Reference 346.
352. A. Schejter and I. Aviram, *J. Biol. Chem.*, 245, 1552 (1970).
353. L. S. Kaminsky, V. Miller, and K. Ivanetich, *Biochem. Biophys. Res. Commun.*, 48, 1609 (1972).
354. P. Edman and G. Begg, *Eur. J. Biochem.*, 1, 80 (1967).
355. S. Moore, in *Chemistry and Biology of Peptides*, J. Meienhofer, Ed., Ann Arbor Science, Ann Arbor, Mich., 1972, p. 629.
356. S. Moore and W. H. Stein, *Science*, 180, 458 (1973).
357. D. Roach and C. W. Gehrke, *J. Chromatogr.*, 52, 393 (1970).
358. F. Sanger and E. O. P. Thompson, *Biochim. Biophys. Acta*, 71, 468 (1963).
359. W. F. Benisek, M. A. Raftery, and R. D. Cole, *Biochemistry*, 6, 3780 (1967).
360. C. H. W. Hirs, *Methods Enzymol.*, 11, 59, 197 (1967).
361. C. H. W. Hirs, *Methods Enzymol.*, 11, 199 (1967).
362. I. Mirakoshi, T. Kuramoto, J. Haginawa, and L. Fowden, *Biochem. Biophys. Res. Commun.*, 41, 1009 (1970).
363. M. A. Raftery and R. D. Cole, *Biochem. Biophys. Res. Commun.*, 10, 467 (1963).
364. A. S. Inglis and T.-Y. Liu, *J. Biol. Chem.*, 245, 112 (1970).
365. T. W. Goodwin and R. A. Morton, *Biochem. J.*, 40, 628 (1946).
366. T. E. Barman and D. E. Koshland, Jr., *J. Biol. Chem.*, 242, 5771 (1967).
367. J. R. Spies and D. C. Chambers, *Anal. Chem.*, 21, 1249 (1949).
368. R. Knox, G. O. Kohler, R. Palter, and H. G. Walker, *Anal. Biochem.*, 36, 136 (1970).

369. H. Matsubara and R. H. Sasaki, *Biochem. Biophys. Res. Commun.*, 35, 175 (1969).
370. L. C. Gruen and P. W. Nicholls, *Anal. Biochem.*, 47, 348 (1972).
371. T. -Y. Liu and Y. H. Chang, *J. Biol. Chem.*, 246, 2842 (1971).
372. T. E. Hugli and S. Moore, *J. Biol. Chem.*, 247, 2828 (1972).
373. Z. Bohak, *J. Biol. Chem.*, 239, 2878 (1964).
374. S. Moore, *J. Biol. Chem.*, 243, 6281 (1968).
375. P. B. Hamilton, *Methods Enzymol.*, 11, 15 (1967).
376. E. Hare, *Federation Proc.*, 25, 709 (1966).
377. S. Moore, personal communication.
378. R. Sterner, unpublished observations.
379. S. Udenfriend, S. Stein, P. Bohlen, and W. Dairman, in *Chemistry and Biology of Peptides*, J. Meienhofer, Ed., Ann Arbor Science, Ann Arbor, Mich., 1972, p. 655.
380. A. M. Felix and G. Turkelsen, *Arch Biochem. Biophys.*, 155, 177 (1973).
381. M. Weigele, S. DeBernardo, and W. Leimgruber, *Biochem. Biophys. Res. Commun.*, 50, 352 (1973).
382. J. M. Manning, A. Marglin, and S. Moore, in *Progress in Peptide Research II*, S. Lande, Ed., New York, Gordon and Breach, 1972, p. 173.
383. J. M. Manning and S. Moore, *J. Biol. Chem.*, 243, 5591 (1968).
384. J. M. Manning, *Methods Enzymol.*, 25, 9 (1972).
385. H. Lindley, *Nature*, 178, 647 (1956).
386. J. Houmard, and G. R. Drapeau, *Proc. Nat. Acad. Sci. U.S.*, 69, 3506 (1972).
387. M. Wingard, G. Matsueda, and R. S. Wolfe, *J. Bacteriol.*, 112, 940 (1972).
388. H. S. Kingdon, C. Noyes, A. Lahiri, and R. L. Heinrikson, *J. Biol. Chem.*, 247, 7923 (1972).
389. V. Mutt, S. Magnusson, J. Jorpes, and E. Dahl, *Biochemistry*, 4, 2358 (1965).
390. V. Mutt and J. Jorpes, *Eur. J. Biochem.*, 6, 152 (1968).
391. M. J. Weinstein and R. F. Doolittle, *Biochim. Biophys. Acta*, 258, 577 (1972).
392. E. Gross and B. Witkop, *J. Biol. Chem.*, 237, 1856 (1962).
393. B. Witkop, *Science*, 162, 318 (1968).
394. I. Koichi and T. Ando, *Methods Enzymol.*, 11, 263 (1967).
395. R. L. Heinrikson, unpublished observations.
396. W. Awad and P. E. Wilcox, *Biochem. Biophys. Res. Commun.*, 17, 709 (1964).
397. N. Catsimpoolas and J. L. Wood, *J. Biol. Chem.*, 241, 1790 (1966).
398. Y. Degani, H. Neumann, and A. Patchornik, *J. Am. Chem. Soc.*, 92, 6969 (1970).
399. R. F. Doolittle, *Methods Enzymol.*, 25B, 231 (1972).
400. M. Z. Atassi, *Arch. Biochem. Biophys.*, 120, 56 (1967).
401. A. Patchornik, W. B. Lawson, E. Gross, and B. Witkop, *J. Am. Chem. Soc.*, 82, 5923 (1960).

402. G. S. Omenn, A. Fontana, and C. B. Anfinsen, *J. Biol. Chem.*, 245, 1897 (1970).
403. Y. Burstein and A. Patchornik, *Biochemistry*, 11, 4641 (1972).
404. J. Schultz, *Methods Enzymol.*, 11, 255 (1967).
405. P. Bornstein and G. Balian, *J. Biol. Chem.*, 245, 4854 (1970).
406. K. Titani, M. A. Hermodson, L. H. Ericsson, K. A. Walsh, and H. Neurath, *Nature New Biol.*, 238, 35 (1972).
407. J. Russell and R. L. Heinrikson, unpublished observations.
408. D. Piskiewics, M. Landon, and E. L. Smith, *Biochem. Biophys. Res. Commun.*, 40, 1173 (1970).
409. R. Sterner, C. Noyes, and R. L. Heinrikson, *Biochemistry*, 13, 91 (1974).
410. R. Sterner and R. L. Heinrickson, unpublished observations.
411. P. P. Fietzek and K. Kühn, *Fortshr. Chem. Forsch.*, 29, 1 (1972).
412. F. W. Putnam, *Fractions* (Beckman Instruments), No. 1, 1973, p. 1.
413. G. R. Stark, *Methods Enzymol.*, 25B, 369 (1972).
414. H. Nau, J. A. Kelley, and K. Biemann, *J. Am. Chem. Soc.*, 95, 7162 (1973).
415. H. -J. Förster, J. A. Kelley, H. Nau, and K. Biemann, in *Chemistry and Biology of Peptides*, J. Meienhofer, Ed., Ann Arbor Science, Ann Arbor, Mich., 1972, p. 679.
416. P. A. Leclerq, P. A. White, K. Hägele, and D. M. Desiderio, in *Chemistry and Biology of Peptides*, J. Meienhofer, Ed., Ann Arbor Science, Ann Arbor, Mich., 1972, p. 687.
417. P. Edman, *Acta Chem. Scand.*, 4, 283 (1950).
418. G. A. Mross, Jr., PhD. Thesis, University of California, San Diego, 1971.
419. G. A. Mross and R. F. Doolittle, *Federation Proc.*, 30, 1241 (1971).
420. M. Bergmann and A. Miekeley, *Ann. Chem.*, 458, 40 (1927).
421. E. Abderhalden and H. Brockmann, *Biochem. Z.*, 225, 386 (1930).
422. P. Edman, *Proc. Roy. Australian Chem. Inst.*, 24, 434 (1957).
423. P. Edman, *Ann. N. Y. Acad. Sci.*, 88, 602 (1960).
424. B. Blombäck, M. Blombäck, P. Edman, and I. B. Hesse, *Biochim. Biophys. Acta*, 115, 371 (1966).
425. P. Edman, in *Protein Sequence Determination*, S. B. Needleman, Ed., Springer-Verlag, Berlin, 1970, p. 211.
426. J. D. Peterson, S. Nehrlich, P. E. Oyer, and D. F. Steiner, *J. Biol. Chem.*, 247, 4866 (1972).
427. M. D. Waterfield, C. Corbett, and E. Haber, *Anal. Biochem.*, 38, 475 (1970).
428. H. D. Niall, *Fractions*, (Beckman Instruments), No. 2, 1969, p. 1.
429. R. A. Laursen, *J. Am. Chem. Soc.*, 88, 5344 (1966).
430. R. A. Laursen, *Eur. J. Biochem.*, 20, 89 (1971).
431. R. A. Laursen, *Methods Enzymol.*, 25B, 344 (1972).
432. H. D. Niall, H. Keutmann, R. Sauer, M. Hogan, B. Dawson, G. Aurbach, and J. Potts, Jr., *Hoppe-Seyler's Z. Physiol. Chem.*, 351, 1586 (1970).
433. G. Braunitzer, B. Schrank, and A. Ruhfus, *Hoppe-Seyler's Z. Physiol. Chem.*, 351, 1589 (1970).

434. J. A. Foster, E. Bruenger, C. L. Hu, K. Albertson, and C. Franzblau, *Biochem. Biophys. Res. Commun.*, 53, 70 (1973).
435. W. Konigsberg, *Methods Enzymol.*, 25B, 326 (1972).
436. W. R. Gray, *Methods Enzymol.*, 25B, 333 (1972).
437. K. R. Woods and K. -T. Wang, *Biochim. Biophys. Acta*, 133, 369 (1967).
438. W. R. Gray and B. S. Hartley, *Biochem. J.*, 89, 59p, 379 (1963).
439. J. -O. Jeppsson and J. Sjöquist, *Anal. Biochem.*, 18, 264 (1967).
440. R. A. Laursen, *Biochem. Biophys. Res. Commun.*, 37, 663 (1969).
441. J. J. Pisano and T. J. Bronzert, *J. Biol. Chem.*, 244, 5597 (1969).
442. M. J. Gordon, *In Sequence* (Beckman Instruments), March, 1971.
443. O. Smithies, D. Gibson, E. M. Fanning, R. M. Goodfliesh, J. G. Gilman, and D. L. Ballantyne, *Biochemistry*, 10, 4912 (1971).
444. D. Gibson and P. J. Anderson, *Biochem. Biophys. Res. Commun.*, 49, 453 (1972).
445. W. M. Fitch and E. Margoliash, *Science*, 155, 279 (1967).
446. K. A. Walsh, *Fractions*, (Beckman Instruments), No. 2, 1972, p. 00.
447. C. A. Woolfolk, B. M. Shapiro, and E. R. Stadtman, *Arch. Biochem. Biophys.*, 116, 177 (1966).
448. R. Walter, D. H. Schlesinger, I. L. Schwartz, and J. D. Capra, *Biochem. Biophys. Res. Commun.*, 44, 293 (1971).
449. H. B. Brewer, Jr., and R. Ronan, *Proc. Nat. Acad. Sci. U.S.*, 67, 1862 (1970).
450. W. F. Brandt and C. Von Holt, *FEBS Letters*, 23, 357 (1972).
451. F. F. Richards and R. E. Lovins, *Methods Enzymol.*, 25B, 314 (1972).
452. R. P. Ambler, *Methods Enzymol.*, 11, 155 (1967).
453. R. Hsu and R. L. Heinrikson, unpublished observations.
454. C. Edelstein, C. T. Lim., and A. M. Scanu, *J. Biol. Chem.*, 248, 7653 (1973).
455. A. Light, *Methods Enzymol.*, 25B, 253 (1972).
456. G. Pfleiderer, *Methods Enzymol.*, 19, 514 (1970).
457. J. K. McDonald, P. X. Callahan, R. E. Smith, and S. Ellis, in *Tissue Proteinases: Enzymology and Biology*, A. J. Barrett and J. T. Dingle, Eds., North Holland, Amsterdam, 1971, p. 69.
458. J. K. McDonald, P. X. Callahan, B. B. Zeitman, and S. Ellis, *J. Biol. Chem.*, 244, 6199 (1969).
459. J. K. McDonald, P. X. Callahan, and S. Ellis, *Methods Enzymol.*, 25B, 272 (1972).
460. R. M. Metrione, A. G. Neues, and J. S. Fruton, *Biochemistry*, 5, 1597 (1966).
461. H. H. Tallan, M. E. Jones, and J. S. Fruton, *J. Biol. Chem.*, 194, 793 (1952).
462. I. M. Voynick and J. S. Fruton, *Biochemistry*, 7, 40 (1968).
463. P. X. Callahan, J. K. McDonald, and S. Ellis, *Methods Enzymol.*, 25B, 282 (1972).
464. R. P. Ambler, *Methods Enzymol.*, 25B, 262 (1972).

465. K. Titani, K. Narita, and K. Okunuki, *J. Biochem.* (Tokyo), 51, 350 (1962).
466. R. L. Heinrikson and H. S. Kingdon, *J. Biol. Chem.*, 246, 1099 (1971).
467. Y. D. Halsey and J. Neurath, *J. Biol. Chem.*, 217, 247 (1955).
468. H. Zuber, *Nature*, 201, 613 (1964).
469. W. F. Carey and J. R. E. Wells, *J. Biol. Chem.*, 247, 5573 (1972).
470. R. Hayashi, S. Moore, and W. H. Stein, *J. Biol. Chem.*, 248, 2296 (1973).
471. L. Polgár and M. L. Bender, *Biochemistry*, 8, 136 (1969).
472. K. E. Neet, A. Narci, and D. E. Koshland, *J. Biol. Chem.*, 243, 6392 (1968).
473. H. Holzer and W. Duntze, *Ann. Rev. Biochem.*, 40, 345 (1971).
474. C. Rochat, H. Rochat, and P. Edman, *Anal. Biochem.*, 37, 259 (1970).

AUTHOR INDEX

Numbers in parentheses are reference numbers and show that an author's work is referred to although his name is not mentioned in the text. The numbers in *italics* indicate the pages on which full references appear.

251

SUBJECT INDEX

CUMULATIVE INDEX
VOLUMES 1, 2, AND 3